Enhancing Me

Enhancing Me

The Hope and the Hype of Human Enhancement

Pete Moore

eat, drink, talk science

Copyright © 2008   Pete Moore
Published by   John Wiley & Sons Ltd, The Atrium, Southern Gate, Chichester,
                    West Sussex PO19 8SQ, England
                    Telephone (+44) 1243 779777

Email (for orders and customer service enquiries): cs-books@wiley.co.uk
Visit our Home Page on www.wileyeurope.com or www.wiley.com

Other Wiley Editorial Offices

John Wiley & Sons Inc., 111 River Street, Hoboken, NJ 07030, USA

Jossey-Bass, 989 Market Street, San Francisco, CA 94103-1741, USA

Wiley-VCH Verlag GmbH, Boschstr. 12, D-69469 Weinheim, Germany

John Wiley & Sons Australia Ltd, 42 McDougall Street, Milton, Queensland 4064, Australia

John Wiley & Sons (Asia) Pte Ltd, 2 Clementi Loop #02-01, Jin Xing Distripark, Singapore 129809

John Wiley & Sons Ltd, 6045 Freemont Blvd, Mississauga, Ontario L5R 4J3, Canada

Wiley also publishes its books in a variety of electronic formats. Some content that appears in print may not be available in electronic books.

British Library Cataloguing in Publication Data
A catalogue record for this book is available from the British Library

ISBN 978-0-470-72409-5
Typeset in 9.5 on 14 pt SM DIN by SNP Best-set Typesetter Ltd., Hong Kong
Printed and bound by Printer Trento in Italy

Contents

Introduction

In 1984 I was fit. I was 22 and supposedly studying a curious degree in farm animal nutrition at the University of Reading, but in reality I spent much of my time training. I was a member of the University first VIII rowing crew, and we were keen and trained ferociously. Winning was key, and there was no honour in second place.

At the same time I was doing the student bit of forming views on the purpose of life and the reasons why we were sure that all generations before us had blown it, of trying to cram work into the smallest possible amount of space to leave maximum opportunity for engagement with others. I'd joined up with a small group of like-minded characters and created a small street theatre group – we thought we were good. The early mornings were frequently followed by as little as 4 or 5 hours sleep. I was alive; really alive. And I knew that if I pushed hard, I could compete with the best.

Then I became ill – the over-stretched elastic band snapped. I'd reached my limit and kept going. I'd not only burned the candle at both ends, but cut it in half and lit all four wicks. It was years before I felt fully well again even though doctors could find nothing specifically wrong with me.

What would I have done to have had an extra 10% of stamina, and an additional 5% of strength, an ability to absorb course work while I slept and to hold it in a database-like memory that enabled me to access it at will and integrate it so that I could create novel associations and ideas? What would I have done if the Student Union shop, alongside its counter of singly wrapped slices of cheese, half-pint bottles of milk and cheap snacks, had had a rack of enhancements for sale? How about a packet of pills that would enable me to work for seven days without needing to sleep? That would have had a deep attraction near to essay deadlines or exam months. How about a spray of genetically modified viruses that, once squirted into the nose, would break through into the blood stream and invade muscle cells, supplying them with the genetic information required to boost strength or stamina?

How about picking up a form to sign up for the next implant session at the University Health Centre? The choice is impressive – maybe intriguing. If I'd signed up on an arts course I could have been tempted to add new sensory options such as magnets or sonar that add other senses to the five I was born with, giving new modes of insight to my work. Alternatively, I could have got on the waiting list for the brain spike that would enable me to tune my mood. No need then for alcohol or other chemical mood modifiers; simply turn the dial left and chill. Turn right and enhance my ability to concentrate on a written assignment. Push a button as I meet my girlfriend and let the evening have a previously inconceivable warmth.

And what of life now? Mid-life and mid-way through losing my hair. What would I like to see on the rack in my local enhancement store? How about the chance of living a little longer – 80 seems a lot closer than it did back in my student days. How about regaining my ability to run, push and lift, or returning my eyesight to the point that varifocal lenses are more than a few weeks away? How about restoring my ability to pick up new ideas and techniques so that I can beat my kids on the PlayStation?

The people chasing hardest after these goals are collectively called transhumanists. This does not mean that they claim to have done anything to actively enhance their bodies, but refers to their aspiration, their desire to see technologies start by fixing biological failure and then move on to create whole new abilities. These could be a capability to transfer ideas directly between people's minds, to 'see' using radar, or to live a thousand years. In terms of numbers, they are a select crowd, but their impact on thinking and policy making is significant. The starting point is an underlying belief that there is nothing particularly clever or good about being human. Nick Bostrom, one of the movement's founders describes transhumanism as 'the intellectual and cultural movement that affirms the possibility and desirability of fundamentally improving the human condition through applied reason, especially by developing and making widely available technologies to eliminate aging and to greatly enhance human intellectual, physical and psychological capacities.'[1] The aim is to look less at what we are as humans, but at 'what we have the potential to become . . . to improve ourselves.'[2]

One problem in trying to look sensibly at this area of endeavour is the need to separate the here, the hope and the hype; to work out what we can do at the moment, what is genuinely likely to be achieved in the next decade, and what is in reality best packaged away between the covers of science fiction novels.

A key to transhumanism is the idea that we can create an enhanced human, but here we strike a problem. What exactly do we mean by enhancement? The first person I interviewed for this book was Anders Sandberg, a research fellow at the Uehiro Centre[3] in Oxford and working for an EU project called ENHANCE that looks at the social and ethical impact of a range of enhancements including mood, physical and cognition enhancements. He's also a research associate at the Future of Humanity Institute. But despite his in-depth knowledge of the area, he struggled when asked to define enhancement. 'Philosophers have been having endless debates about the definition of enhancement. I think there are roughly four approaches to defining enhancement.'

[1] www.transhumanism.org/resources/faq.html
[2] www.transhumanism.org/resources/faq.html
[3] http://www.practicalethics.ox.ac.uk/

The first, he suggests, is to avoid anything as restrictive as a definition, but instead to point out a few examples of enhancements. 'People in general seem to understand that approach', he says.

The second approach again is not to define enhancement, but to talk about its great importance. Interestingly, Sandberg points out that this strategy is used both by transhumanists arguing for enhancement, and by bioconservatives who are keen to see it stopped. 'It's so amusing to see both sides essentially use the same strategy, because from each of their perspectives this is about the deep human values. It's terribly important', smiles Sandberg.

The third way forward does create a defintion, and that is to say that enhance-ments are medical interventions that are not treatments. The problem here is that no one can agree on what constitutes valid medicine, or what is an illness. So depending on your definition of illness, health and medicine, you will reach a different definition of enhancement.

'I come from computer science', says Sandberg, as he introduces the fourth definition, 'So I'm trying to simplify the approach. For me, enhancement is something that improves the function of some core capacity of the mind.' It could be an increased ability to store and recall experiences and information; to think about them, retain them and get the relevant information back when we want it. I wonder, though, whether that definition is too tight to be useful.

Across the room in Sandberg's Oxford office is Professor Julian Savulescu. He has yet another defintion of enhancement. For him, an enhancement is 'any change in the biology or psychology of a person which increases the chances of leading a good life in circumstances C.'[4] He says that 'to enhance is to increase

[4] J. Savulescu, Justice, fairness, and enhancement. Annals of the New York Academy of Sciences. 2006; 1093(1): 321–338.

in value . . . to increase the value of a person's life.' He also refers to a 'narrow definition of enhancement: any change in the biology or psychology of a person which increases species-typical normal functioning above some statistically defined level.' In other words, you are enhanced not if some technological fix enables you to do something that no other human can, but instead helps you to do something that is impossible for the vast majority of human beings. An instant problem here, however, is in determining any statistical 'norm'. It would also mean that world record-holding atheletes, by definition, should be considered to be enhanced, even if this enhancement has occurred through genetic mutation and extreme training.

It strikes me that one of the ways of trying to make sense of all these definitions is going to be to take a look at what is going on, and who is calling for it, and then see which of these definitions has most to offer.

The best way of finding out what is really possible is to talk with people who have done it already. Much to the transhumanists' annoyance, most of us sit down and accept our lot with simple resignation. Last time I called in at my doctor and described my aliment his reply was prefaced with the four-word put-in-your-place of 'well, at your age . . .'. Little hint there of a 'let's beat biology' mentality cutting in.

There are, however, a few transhumanists who claim to have had enough 'get up and go' to have 'got up and gone' and moved on from simply talking about the possibility of enhancing their bodies, and started to build in new features. Through this book we will meet some of them and look at their rationales and their responses to the experience.

Taking this approach, however, kicks up a problem. There are not many such people to talk to. Or are there? It all depends on how you define 'enhanced'. Technology is frequently added to people, or inserted into them, to solve medically diagnosed deficiencies. Which of these are therapies and which are enhancements? If you wear glasses are you enhanced or simply enabled to live normally? What about putting on night vision goggles? Now you have a new capability, a novel ability to navigate at night without the need for light that could give away your location. How about a heart that continues to send your blood coursing through your veins long after the organ you were born with ceased to beat? Is it an enhancement if you have a bolt-on technology that lifts you from a state of deep disability to one of partial ability – less than fully normal, but much more than you had before?

My solution to this conundrum is to include technological fixes to personal problems within the group of issues I'll explore and discuss the techniques with inventors. I can then talk to the people who rely on them and let them say whether they feel enhanced. The history of human integration with technology is that it is almost invariably tried out on those with conditions that are so disabling they are willing to try almost anything to alleviate their situation. Their plight also makes ethical committees more likely to grant permission for them to take part in a trial of some new device.

One consequence of this is that we need to be careful when we make rash assumptions about who the early adopters of human enhancement are likely to be. It strikes me that they may well be very different from those who snap up the latest High Street gadgets. They are more likely to be people struggling on the edge of society, rather than the life-pushing executives, or the experience-seeking teens and twenty-somethings.

I feel I need to show my hand at the beginning of this book, to let you know where I'm coming from. Unlike most books written about transhumanism and enhancement, I'm neither a passionate advocate, nor am I a dyed-in-the-wool critic – I'm a curious questioner, an enthusiastic sceptic. I quite like being human, and am not sure that I want to become something else. I struggle to make use of all the senses I have, without having a constant desire to add new ones. I also have a somewhat sensitive hypeoscope – a built-in sensor that sounds a warning when in the presence of an argument that is slipping seamlessly from the implementable to the impossible without pausing for breath. I'm setting out on the journey of writing this book as a sceptic looking for evidence to convince me that human enhancement is here, needs to be taken seriously and should start to inform policy-making in the twenty-first century.

I also want to take a look at the underlying desires that drive the enhancement movement and see which ones of these are likely to be widely held, or whether they are the wishes of a fanatic few. The distinction is important as we assess the extent to which we would like an enhancement-driven agenda to shape our health, education and social systems. Quite often policy change can be driven by small pressure groups of well-coordinated activists. To recognise that is not to add a value judgement – most people would agree with the abolition of slavery, a policy change being celebrated in the UK as I write this book, and arguably brought about by a small group of vigorous campaigners. It is, however, a good idea to check the ideologies of such groups early, to see if they are likely to lead society in a beneficial direction, even if the majority then sit back and let the minority do the work of invoking the change.

As part of this, it will be important to look at the business opportunities unleashed by enhancement technologies. If an idea is going to reach a mass population, it will need someone to supply it. A wonderful technology that cannot reach the market will have great difficulty impacting the lives of more than a few. Conversely, even weak enhancements that are easy to market could potentially invoke considerable change if widely adopted. This will inevitably highlight the ethical difficulties faced by the pharmaceutical industry, among others, who are often accused of profiting at the expense of distress.

In previous books, I've written the introduction last. It has normally served as a means of summarising my destination and enabling the reader to discover the contents without being troubled by the task of reading the text. This time I've written the introduction first. This book is a genuine journey of enquiry and I trust that it will be one of discovery. It is researched and written over the summer of 2007 and as such serves as a snapshot of the situation, to meet people who are dreaming, devising, delivering and using technology that could shape all of our futures. Snapshots have to be created quickly, so it's going to be a rather breakneck journey, and I had the good fortune to bump into Heather Bradshaw at the outset, who became an invaluable assistant and researcher. Her thoughts and knowledge of the area gave me a tremendous head-start, as well as useful food for thought.

I'm currently on a cross-country train on the way to interview one of my witnesses – I invite you to travel with me.

Pete Moore
5 August 2007

I Longer Than Life

When the kids from *Fame* sang 'I want to live forever', I don't think they really meant it. They might have sung with gusto but, as they leapt across the stage in tight jeans and T-shirts, their thoughts were less about existing in a thousand years' time than desiring the longevity of their name. The great wannabe hope is that you will be recognised for your talent and that years after you have been placed six-foot under, they will remember your name.

An alternative hope is the desire to live after death. It's an ancient goal and one found in most cultures. It often comes with a desire that this next life will be better than the current one – think pyramids or armies of Chinese clay soldiers or iron-age burial sites with useful and costly tools laid alongside corpses.

But, in the view of many, all of these routes to immortality have a key problem. The after life will only ever cut in after you have gone. To live for ever in the most literal sense would require that you don't die in the first place.

To stay here for ever, or at least for longer, will mean driving ageing out of the system, and various proponents have suggested a number of options. I went in search of people who are either doing, attempting to do or thinking about these things.

1 Techno Fixes

The rain on 20 July 2007 made history as it flooded the UK midlands – not a good day to book an interview in a suburb of Birmingham and have to journey through the thick of it to get there. Still, the machinery that enables our busy lives was good enough to get me there, even though I did get soaked in the final 50-metre dash I had to make from the car to my interviewee's front door.

The aim of this trip was to meet Peter Houghton. The interesting thing about Peter is that he should really be dead – in fact, in one important way he was dead for a year or so. Diane, Peter's wife, welcomed me and took me upstairs to his room. Peter, whom the Guinness Book of Records states is the longest living person in the world with any kind of artificial heart, sat in front of his computer. Behind him shelves rose to the ceiling gently cluttered with books and magazines, and holding a few small Catholic icons.

The only sign that this was the room of a man who should have been dead was the stack of plastic boxes of medications in the far corner – few people would consider the large battery charger on top of a filing cabinet as an essential part of his medical paraphernalia.

A Bad Bout of 'Flu

'I'm a retired psychotherapist', Peter told me when I had got settled. 'I'm 68 years of age, and I've got a Jarvik 2000 artificial heart pump, known as a left ventricular assist device.'

Peter's problems started in 1995 when he had a very bad bout of 'flu, finishing with a heart attack. He was treated with drugs that managed to control the symptoms reasonably well for a while, but they couldn't help entirely. His heart was too damaged.

A healthy heart works by letting blood flow into its four chambers through one-way valves. These valves flop closed when the heart starts to contract – it's the opening and closing that cause the classic 'lub-dud' sound much loved of films and TV hospital dramas. With the valves now closed and the heart's muscles contracting, the blood has no option but to move on through the system. To work properly, therefore, the heart needs its muscles to be fit and the valves to provide a tight seal. Peter's illness had caused the heart muscles to expand, but the valves remained unchanged. As a result the main valve, the mitral valve, no longer fitted properly and was leaking badly. There was no obvious surgical solution.

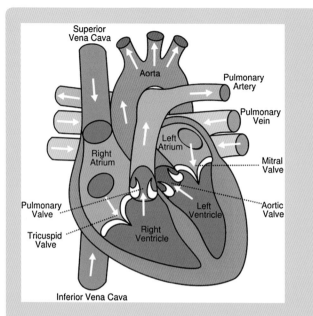

The heart is as complex as it is simple. It is composed of four chambers arranged in two pairs. On the right side is a thinly walled sac, the right atrium. The walls are basically made of muscle, and the sac collects blood as it flows in from the body. At the beginning of a heartbeat, the right atrium contracts and gently presses this blood through a valve into the right ventricle. This has much thicker walls and when it contracts it pumps blood into the artery that supplies the lungs. In the lungs, the blood picks up oxygen and drops off carbon dioxide, before heading back to the left side of the heart and entering the left atrium. From here it is pushed into the left ventricle – the heart's largest and most powerful pumping chamber. When this contracts, blood is forced into the body's main artery, the aorta, which channels it towards all the organs.

By February 2000, Peter's condition had deteriorated so far that he began to plan his funeral and tie up the loose ends of his life. Medicine apparently could do no more. He had very little energy and could hardly walk. He was sick from impending kidney failure and suffered from gout, skin infections and itching. He was losing the function in outlying parts of his body. His end looked very near.

The Jarvik 2000

Then in June he agreed to take part in the medical trial of the Jarvik 2000 rotary heart pump, and had the Jarvik Heart implanted at the John Radcliffe Hospital in Oxford on 20 June 2000.

'So what does the machine actually do? You've still got your heart?' I asked.

'Yes. Well, the main part sits in the left ventricle, and it does the pumping work for the ventricle. The pump collects blood from the ventricle and pushes it through a 2-centimetre-wide tube straight into the aorta. If you listened, if you got a stethoscope and you listened, all you'd hear in my heart was "bzzzzzzzzz".'

This pump is a marvel of miniature engineering and technology. It is a polished titanium-cased axial flow pump that weighs about the same as a small apple (85 grams) and is only as big as a 'C'-sized battery (25 cubic centimetres in volume). Even so it is capable of shifting up to 10 litres of blood a minute. The engineering design is impressive, but the technology capable of creating materials that survive being implanted in the body for years is even smarter. Blood is a highly corrosive fluid and sets about attacking anything it touches, so the materials must be carefully chosen. In addition, there is no point pumping the blood around if the pump damages the fragile blood cells, so the structure of the rotors and chamber is critical.

'So, is your heart beating?' I said, looking at the calm man sitting opposite me on a blue office chair.

'It is now. It didn't for a long time', he replied.

Now I don't know about you, but for me a basic feature of a living human is that he or she has a beating heart. It's a simple part of basic biology. If you went back to mediaeval times and the heart wasn't beating, that would be a reason for putting you six-foot underground. But Peter is sitting, literally as living proof that a beating heart may be a common feature of life, but it isn't required. So long as something is keeping the blood flowing, life can move on.

'One of the questions that several people have asked me is; are you dead really?' He chuckled, 'I'm never quite sure how to answer it.'

The Jarvik 2000 rotary heart pump combines miniaturisation with durability. While the power source is outside, the pump is very much located in the body.

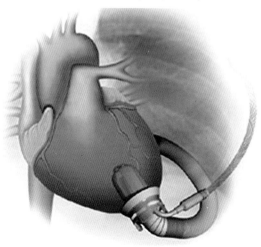

Peter's pump had been in place for seven years, and it is only over the last couple that his heart has started to beat again, generating a weak pulse. I felt his wrist to check and it was there, but strangely faint.

Peter looks a little tired and pale but otherwise quite alive, despite the curly 'telephone cable' coming out from behind his ear. The cable runs into a grey camera bag which sits like a penguin's egg between his feet. The implant is reliable, but Peter depends on electricity for his blood flow. The bag on the floor contains some paperback book-sized rechargeable batteries, and there is a backup battery built into the implant that will enable it to operate for up to 20 minutes – enough time to change the main batteries over. Cables pose their own engineering challenge. If they bend they are prone to fatigue and snap over time – giving a panic-filled few minutes. Peter's fingers are too affected by rheumatism to change the cable himself, so he relies on quick attention from Diane. Thankfully,

Peter's power cable fixes to the side of his head.

the internal cable is more protected and has never failed – if it did, that would probably be fatal.

While the pump is inside, the external batteries are an interesting extension to his body. He can never be more than a cable length from them, and as such the batteries and bag are part of the picture. As with any part of your body, Peter is pretty wary about strangers touching it. It's an understandable concern, particularly as in the first year he had the pump, a young man snatched the bag and ran off, pulling the plug out from Peter's head. 'It was in front of Marks & Spencer', explained Peter. 'I'd just bought a pair of trousers. The bag started to bleep like hell and the boy dropped it, and a little old man with a Zimmer picked it up. He said, "I'm glad someone's in more trouble than me!"'

Peter has a habit of staring at the floor, as though lifting his head is too much like hard work. Yet when he does look up it's with a quick, sharp movement. His face and particularly his eyes are alive with intelligence and a measure of wisdom, not to mention a wicked sense of humour.

Neanderthal Cyborg

So Peter Houghton continues to think, talk and best of all to laugh even though his circulatory system is definitely not operating in the normal way for the species. But what is he? Is he an enhanced transhuman, a lucky heart patient or some kind of freak?

'I'm still just trying to work out whether you see yourself as a biological mechanical amalgam now, or . . .'

'. . . am I a cyborg, do you mean? I think I am a cyborg, yes, to some degree, and I'm a Neanderthal cyborg.'

❝ I think I am a **cyborg**, **yes**, to **some** degree, and I'm a Neanderthal cyborg **❞**

Peter states that he is a cyborg because he is kept alive by mechanics, because his implanted pump enables nutrients, hormones, gasses and so many of life's basic ingredients to keep flowing. The Neanderthal part of his self-assessment comes from the fact that while he would like to leap buildings, become one with his computer and break new records of achievement, the reality is that with current technology he is struggling to survive – anything more substantially successful, he feels, will be for distant generations.

Inside Out

I absent-mindedly took my glasses off and while polishing the lenses tried to work out whether I would feel differently about myself if these lenses had been implanted inside my eye. I don't feel that the glasses are 'me', but I am lost without them. At the same time, though, others who know me do see the glasses as part of who I am. My children when young were even somewhat alarmed when I got a new pair – as far as they were concerned, I had changed.

I wondered if Peter would feel differently about his pump if it were external. The point here is that when some people talk about human enhancement, they seem to give an importance to what's placed within the body and are prone to dismiss anything that is positioned without. To an extent, this heart pump is an ideal candidate to see whether this is indeed important, as there are people with tubes that push through their chest walls and connect to pumps on moveable trolleys. It turned out that Peter had met several of them on trips to the States. Peter had in fact been surprised by how much they had had in common, with the only real difference being that other patients are envious because they felt that the internalisation of the pump meant much less was able to go wrong and would therefore give a much better chance of survival.

Could it be that the location of the unit, inside or out, is less of an issue than the function it performs, or the permanence of its use? If your heart were on the outside, it would still be very much 'you' because without it you would cease to live.

Less Flow than Normal

What sort of a health does Peter Houghton have now? In physiological terms, the left ventricular valve of a fit young healthy person pumps about 80% of the freshly oxygen-loaded blood that reaches it from the lungs off around the arteries. Only 20% would remain in the heart cavity. As you age, this proportion changes and a person of average fitness in his or her middle years might only be sending 45% round the body with each beat.

Peter explained that before he had the Jarvik 2000 fitted, his left ventricular valve was sending only 8% of the blood it received around his body. That really is not enough for the major organs such as the brain, kidneys, liver, stomach and limb muscles to keep functioning. That's why he was barely clinging on to life.

Getting the blood moving again was important, because it is blood flow rather than a beating heart that is critical for life. How much flow does Peter have with the Jarvik 2000 fitted?

It is **blood flow rather** than a beating heart that is **critical for life**

'It's 25% of normal.'

That seems pretty low, but unlike some implants, the Jarvik 2000 does not aim to restore normal flow levels. For most people it is used as an assisting device and not a replacement. The rationale for this is two-fold: firstly, it needs to be as small as possible so that it can fit more neatly inside the chest, while being built to last. Secondly, the Jarvik 2000 aims to take some of the heart's workload, so that the organ is given time to recover from disease. In Peter's case, it is having to work at the top end of its ability.

Advanced or Enhanced?

So Peter sees himself as a cyborg, but is the heart an advanced surgical splint or an enhancement? Does it provide him with any benefits over his original, squashy heart? For example, is it reliable in a way that his natural one wasn't?

'The great thing about the Jarvik heart is that it's proved so robust', Peter said. 'There are several other artificial hearts about that have not proved robust. They fail on various mechanical parts very quickly and you can't replace them quickly enough. But Jarvik's great advantage in the market at the moment is that it has proved so strong and so reliable.'

Certainly, in terms of medical engineering, seven years is a good test of any pump, but how will it fare over 15 or 30? It may be good in engineering terms, but it has a long way to go before it proves it can match a natural one.

But is there anything that Peter can do with his mechanical pump that I can't do with my heart?

'Well', he said. 'I've got a controller, with which you can increase or decrease the speed of the pump. So, say you're climbing a hill or walking up stairs, you can give yourself a little bit extra. And at night, if I left it on the standard setting of "three", I would never fall asleep. I have to lower it to "two", to prevent my brain keeping too active. So to a limited extent I can have some control over it.' He would, however, clearly like greater control over the heart.

But does he really want manual control over it? Most of us don't even realise the way that our hearts vary their output through the day. Is a control dial an advantage, or something that creates an inconvenient chore? If you ask a car manufacturer, they would probably say it is a chore. Just think about the new range of cars with windscreens that detect rain to take away your need to flick a switch. They would be wanting to hand this control over to some automated function really rather quickly – but that seems a long way away.

Also, while the majority of humanity regulates breathing and heart control as a total package, Peter has to manage his heart rate and his breathing separately. That's quite a challenge.

Then Peter introduced a new twist.

'And I can turn it off', he said in a flat tone.

It was a shocking thought. Simply pulling the plug from his head-mounted socket, or deciding not to replace empty batteries could end his life. We all know that we can take actions that will end our lives if we so wish, but the 2-millimeter-wide cable reaching from bag to head seemed to take on an even greater importance. Peter, however, is reassured by the fact that he has critical control over the technology and can 'get out of it all' if he feels the need. He knows that if he becomes too weary of living with substandard blood flow, he can flick a switch and quietly move on.

Extending Existence Is Not Necessarily Enhancing

Are the drawbacks balanced by the extra capabilities to control and to rely on technology? What does Peter make of the extra capabilities?

'You haven't got that. You haven't got any capabilities that aren't normal', he insisted.

'Well, but there are some', I pressed. 'There are the . . . I think I might like to go to sleep now, let's just slow it down. There are the . . . I think I can do with concentrating a bit harder, let's turn it up; I think I've had enough of it, let's pull the plug. Now I can't actually do any of those in quite such a prescriptive manner. I might decide to go for a walk as a roundabout way of tuning up my concentration. I might have a glass of whisky as a roundabout way of trying to get to sleep on a day when my brain's chasing round. I'd have to do something really quite proactive to draw things to a close quickly. So it is a range of capabilities that you might not want, but that I don't have.'

'Fine', he said crossly, 'but the word "capabilities", that's the one that bothers me. I'm not capable of doing what you do. I understand your point, but in fact, I am a doddery old man with limited access to anything really. It's not my experience . . . I may have certain choices I can make, but choices aren't the same thing as . . . as . . .'

'. . . capabilities?'

'Well, I would say I haven't been enhanced, you know, as simple as that. I just don't feel enhanced. I feel sick a lot of the time and therefore, you know, I don't feel enhanced. And I have to deal with life on that sort of basis, which is very unsatisfactory really', Peter emphasised.

But isn't the fact that he has an extra span of life, isn't that in itself an enhancement? Not for Peter. From his point of view this is something that has just happened, but despite the fact that the pump had kept him going for seven years and could do for as long again, possibly more, he flatly was certain that he was not enhanced.

I was intrigued by this, because I'd just been reading one of Nick Bostrom's papers. It had been mapping out the history of enhancement and claiming that the concept started with the alchemists and their search for the elixir of life. This would imply that a technology that rescued people from the brink of the grave and gave them more than a 10% extension of lifespan was a strong candidate for a claim of being enhancing, and there are many occasions when Bostrom has pointed to Peter as a living example of a prototype enhanced human. But for someone living it out, for Peter, the situation is less exciting – life extension on its own is not all it's cracked up to be. It's not necessarily an enhancement.

Nick Bostrom

Dr Bostrom, a member of the Faculty of Philosophy at the University of Oxford, is currently the Director of the Future of Humanity Institute, part of the James Martin C21 School at the University of Oxford.

Bostrom has a background in physics and computational neuroscience as well as philosophy. Before moving to Oxford, he taught philosophy at Yale University. His research interests include the ethics of human enhancement and consequences of potential future technologies such as artificial intelligence and nanotechnology, as well as foundations of probability theory and global catastrophic risk.

He has published more than 100 articles, and one book entitled *Anthropic Bias*. His writings have been translated into more than 15 languages.

A favourite media boff, Bostrom has given nearly 200 interviews for television, radio and print media.

'Well, it's not at all what it's cracked up to be for *me*, because I can make litlle use of it. I'm too weak and I don't have the strength to do that.'

Value of Life

So Peter doesn't feel enhanced, but that's understandable, he was less than fully well. Does he see this as a medical device, a therapeutic device, an enhancing device, or a pain in the neck?

'All those things', he laughed, 'especially the latter! How do I see it? I mean, . . . first of all, it's an opportunity because it gives me a bit of extra life, if I can answer reasonably the question of what it's for. The only problem is, I haven't really answered *that* question satisfactorily.'

As a practising Catholic, Peter goes to mass or has it brought to his home, and is concerned about living a life that is of service to God. His faith is important to the way he views his situation and is also an important part of his coping strategy. Despite saying that he is not enhanced, he is clearly pleased to be spending longer on earth working out his faith and relating to family and friends.

But, I wondered, has his new heart changed his own sense of identity or affected the way he values life? His answer had elements of yes and no. No, the pump itself hadn't affected the way he sees himself or his life, but the whole process of being so ill, and of sitting day by day with a battery pack at his feet and cable in his head was a constant reminder that he was unwell. His general lack of energy was another clear indication that he was less than fit. The outcome of this was a distinct difficulty in engaging in the life that was going on around him. Family events such as grandchildren being born or critical birthdays came and went with pleasure but not excitement. Each was dampened down by a fear of getting too involved and of consequently enhancing the pain when he did eventually die.

Booking a Legacy

Peter's problem is that while he is embedded in the present, he is cautious of engaging with things that demand planning for the future – a future in which he may no longer be. His brush with death has turned into a prolonged game of brinkmanship, of staring the grim reaper in the face and trying not to blink. And more worrying for Peter is the fact that his family are forced to join him in this process – to share this prolonged epilogue.

Like many, his planning for the future has shifted subtly to making sure he has created a tangible legacy. 'I'd got this sort of idea, you know, what sort of legacy could I leave and I'd got to write a book', he explained.

In fact he has written three: *On Death, Dying and Not Dying* is autobiographical, *The World Within Me: A Personal Journey to Spiritual Understanding* is religious, and one is a work of science fiction.

Perhaps if the Jarvik device had been offered earlier to Peter, before he had cut off so many of the relationships that formed his life, and brought his own story to a close so efficiently, he would feel more positive about the extra years, even though his quality of life is not as good as he obviously wishes. What if a new generation of Jarvik pump delivered a higher blood flow? Would the increased sense of health alter Peter's views? It probably would, but that capability is still a long way off.

Opportunities for Enhancing Health

There is another option that needs thinking about. In theory, we could fit healthy people with Jarvik devices with the aim of extending life – it has been done in two sheep, and they lived longer than most. We could do this well before a person had come to the end of her or his life, knowing that it would prolong the heart's operating life and reduce her or his chances of dying from heart failure or its complications. Such people would face fewer of the psychological issues that Peter, as the pioneer, is concerned with. They might well be enhanced in having a heart-system that was expected to have a lifetime well beyond the human average.

Who, I wonder, in their right mind would want to go through the 14-hour operation with all its risks, side effects and the lengthy convalescence? Would the benefits to a healthy person ever outweigh these costs? Today it is difficult to imagine the costs reducing enough, or the value of the benefit rising enough, to change the outcome of the calculation. Mechanising the heart seems an unattractive way of reaching beyond the average human lifespan.

Moving the 'Average' Goalposts

Yet, by being available for those whose natural lives have been cut short by heart failure, Jarvik devices are part of the whole technological environment that is pushing the average life expectancy, in wealthy countries, up significantly. Each individual case may be therapy and not enhancement, but the result of widespread use would be to change the statistical distribution of human life expectancy. The average age of death would shift higher and the range would narrow.

Look at it at an individual level and it may be difficult to spot any enhancement – certainly Peter is highly reluctant to go down that route – but at the level of the species, average lifespan has been extended. Some people would equate that with 'enhancement', others would say it is progress. As such Peter's pump is doing nothing particularly new; rather it is simply finding another way of enhancing humanity as a whole, like vaccination, sewers and the elimination of dangerous wild animals.

How Much Is You?

Peter finds himself torn between a desire to live long enough to make use of future technologies that could potentially increase his ability to thrive and achieve, and his anxiety that this would only come about by replacing ever-increasing amounts of his body with mechanical devices. His dilemma is that being mechanical doesn't sound deeply attractive. Having a mechanical heart has sent him one step down that line, and begs the question of how much you can replace and remain 'you'.

For Peter, the route to answering the question is to ask what the soul is, what the part of him is that will survive death. As a Catholic he is convinced that this non-physical element of existence is the real him, the part that matters. He is not prepared to limit his thinking to saying that the feature that is 'essentially you' is just a product of the mind reacting with the physical body, but adds to this mix that each person is someone whom God made with his or her own uniqueness. Each person has a spiritual dimension as well as a physical one.

In this assessment, the body is important because the abilities and restrictions it brings shape who you are – but it is not everything. In most religious philosophies, relationships are key to determining individual identity and value. As with all forms of Christianity, Catholicism focuses on a person's relationship with God, with other humans and with the rest of the created universe. In this assessment, the enhancements that most powerfully affect who you are would be those that alter your ability to build these enlivening relationships.

Maybe this is why Peter seems less than enchanted by his techno-heart. While being thrilled to have it, and excited to be a world record-breaking survivor, the mechanism hasn't given him an enhanced ability to relate to others. In fact the sub-normal level of function leaves relationships extended in time, but in many ways, depleted in depth.

Moving On

Outside the rain was heavier than ever and the journey home was clearly not going to be easy. It had been a pleasure to meet Peter and I hoped I would come up with another excuse to bump into him again.

Sadly I was not able to meet Peter again. Through the summer his health deteriorated and he died on 25 November at Birmingham's Selly Oak Hospital. Hearing the news made me realise just how much of a privilege it had been to meet him.

I sprinted through standing water to the car and set off through a scene of surprising devastation south of Birmingham; car roofs sticking out of brown lakes of floodwater here, water pouring through letterboxes of houses there, muddy torrents deep enough to defeat most cars flowing across roads in hundred-yard-long sections. Only the few people who had off-road equipped Land Rovers had any hope of getting home that night, but even they struggled, and some floundered.

Struggling through Henley-in-Arden showed the ease with which machines could become overwhelmed by natural forces.

It seems that much of our life is supported by machinery, most of which we can climb out of and discard if life's floods get too deep. But on a good day, it enables us to travel further, to explore areas of the world that humans would never penetrate unaided, to simply cram more into life. It can support us when we are too weak to support ourselves. As such, technology does enhance our lives, but I'm not sure so far that I've found evidence that it can create enhanced humans.

2 Maintain and Repair

I'd gone to see Peter because he was often held up as one of the examples of what an enhanced future could soon be like. It had been fascinating to meet a medical marvel whose life was drawn out beyond the point of near fatal damage. But from the point of view of human enhancement, he wasn't exactly a triumph. I was therefore keen to investigate other hopes for extending life, technologies that could give us many more years than the standard three-score-and-10 expectancy in a developed country. I went in search of another big noise in the enhancement world.

On one of the rare sunny days in 2007, I took the train to London. This time my quarry was Aubrey David Nicholas Jasper de Grey. His Wikipedia entry starts by saying that he 'is a biomedical gerontologist who lives in the city of Cambridge, UK. He is working to expedite the development of a cure for human ageing, a medical goal he refers to as engineered negligible senescence.' Hopefully, he will talk to me using shorter words, but my reason for tracking him down is that he has a reputation for stimulating thinking and work that aims to end ageing – if he can deliver, then that would be a highly sought-after enhancement.

Aubrey de Grey's dress sense is as flamboyant as his ideas. Both set out to break moulds and place him at the leading edge of a new world.

We met a few hundred metres from the British Library at the Skinner's Arms, a traditional London pub with an oak bar, gold lettering on mirrors, and a sense of place and history that cannot be replicated by any quick designer make-over. De Grey was tucked in a corner, pint in one hand, book in the other as I introduced myself. We'd never met, but he was easy to identify. I bought myself a large orange juice, turned on my recorder and started . . .

De Grey's interest in combating ageing is driven by a belief that it is a bad thing that a hundred thousand people a day die of ageing and most of them die really horribly. To do something about it he has developed a multi-part scheme that he hopes will combat various aspects of ageing, and has established a US-based charitable foundation that makes money available to people who want to pursue the ideas. He doesn't have a laboratory of his own, preferring to do the theoretical thinking free from the restrictions imposed by learning to perform specific methods and techniques.

His main concern is not so much to extend life, as to eliminate ageing: 'The fact that two-thirds of all deaths worldwide are due to ageing, and perhaps 90 per cent of all deaths in the industrialised world are due to ageing, means that if we were to eliminate ageing, people would live a lot longer on average, unless they chose not to. But the purpose, in my view, is to eliminate involuntary death from ageing, to give people the choice of whether to live to a hundred when they're 99, rather than to have that choice progressively removed from them by their declining health. Indeed, to have the choice to live to a thousand when they're 999', he said.

But the **purpose**, in my view, is to eliminate involuntary death from ageing

At this point I did a quick bit of mental maths. As it stands, the world has around 6.5 billion humans, many of whom are struggling to find enough food and water. If nothing happens to the basic birth and death rate, the current projections suggest that it will rise to 12.0 billion by 2050. The expectation is that, as people get wealthier they have smaller families, and so the birth rate is likely to decline, but even then most people think we will still have in the order of 9 to 10 billion by halfway through this century. Supporting that number is going to be quite a task – supporting the number that would be generated by thousand-year lifespans would be an extreme challenge.

The only way out would be to have a moratorium on breeding. As someone with three children, I find it difficult to believe that life would be as satisfying in a sterile non-breeding community. Forcing it would also infringe what we believe are basic human rights – not good for an enhanced life.

'People often say to me well, hang on, I don't even want to live to a hundred or certainly not a thousand', De Grey commented, 'and I reply, absolutely, I don't have any particular active wish to live to a thousand. I don't even know whether I want to live to a hundred, at this point. I'm only 44 now, and I may think differently when I'm 45. I may definitely think differently when I'm 99, if I'm still biologically only 44. But this issue is to give people progressively more choice, rather than let it just flow away.'

Made Do and Mend

To defeat ageing, De Grey proposes that we set about a program of repair and maintenance he calls SENS (Strategies for Engineered Negligible Senescence). The concept is derived directly from a belief that the human body is a machine. It's a really, really complicated machine, but it is a machine. We know from the teams of enthusiasts who keep steam engines puffing, or from the regular trips to a garage with our cars, that good maintenance can keep mechanical objects going long after other equivalent examples have hit the scrap heap. If you repair damage before it develops too far and causes secondary problems, then you will elongate the machine's life expectancy.

If you **constantly replace bits** of the body that **wear out** – how much of you will be **left**?

The steam engine shows that with enough repair and replacement of worn-out parts, you can keep the thing going almost indefinitely. But this does raise an interesting issue. Steam engines survive because portiors of them are replaced bit by bit. After a hundred or more years of maintenance, much of the original mechanical beast is no longer there. If you can constantly replace bits of the body that wear out – how much of you will be left?

'Now, it'll be a while before we get to that point with humans, but perhaps not all that much of a while', said De Grey. His strategy is to start by figuring out what sorts of damage accumulate in the body over time, and then work out how to reverse and repair those various types of damage.

Don't Worry About Causes

One interesting aspect of this SENS approach is that De Grey believes we no longer need to worry about why the damage occurs; we will simply recognise it and then repair it. This doesn't mean that we can have a reckless disregard for our health and safety, but it shifts the balance of concern. He claims that one of the advantages of this is that it avoids knowing too much about how the body works, or what goes wrong in ageing. If a piece goes wrong, you put a new one in. You don't need to know what it is doing, you just need to make sure that it is present and correct.

If a **piece** goes wrong, you put **a new one in**

'All we have to know is the actual differences between, let's say, a 20-year-old and a 40-year-old, okay, and we have to figure out ways to reverse the differences', he tells me. 'It's not even important which of those differences matter more in terms of causing pathology in late life and which of them matter less. If we fix more-or-less everything that changes, then sure, we may fix one or two things that in practice we didn't need to fix, but no harm done. Ultimately, structure defines function, so if we restore the structure of the human body to something like what it was at a younger age, then we will be restoring the function, and the function includes the remaining healthy lifespan.'

De Grey explained that from the moment we start developing in the womb we accumulate bits and pieces of molecular damage. Initially, this has no effect on how well you function, but there comes a time when the damage is too great to ignore, and bits of the body start to fail. His hope is that there is a mid-life window of opportunity during which you can repeatedly clear out rubbish, much like putting refuse sacks out from the house, so as to continually keep the body in good order.

The problem with chasing after a concept of 'prevention being better than cure' is that to prevent things going wrong in the body, we need to have a deep understanding of how it works. And De Grey is concerned that we are a very long way from that, believing that our ignorance of the way cells work, let alone the way that organisms work, is profound. Science has come a long way, but in reality it has hardly scratched the surface – it is only just beginning to see the incredible complexity within each cell, and has a long journey to travel before it makes sense of it. His repair and maintenance approach therefore does not act like a geriatrician and target pathology, but to intervene at the weak links in the chains of events that lead to ageing.

The more I think about this line of argument, the more worried I get. It seems to me that there is an inconsistency in De Grey's logic here. On the one hand he says that we are very ignorant about what is actually going on inside cells, too ignorant to be successful in tackling it head on. On the other he says that what we need to do is not worry about the processes, but simply replace damaged bits and restore healthy function. What I'm not sure about, though, is how we are going to build the replacement parts unless we can understand the exact way the broken pieces should be working. Even if we are planning to repair existing cells and organs, we need to know what they are like when they are working well. Then we can spot when they go wrong and know what we are trying to restore them to. This will still demand a high level of biological knowledge.

The Seven-Point Challenge

Far from naïve to the scale of the challenge, De Grey still believes that it is a route that has the greatest chance of creating a successful outcome. In a manner similar to taking yourself to a garage for an annual service and road worthiness test, De Grey proposes that we should create a checklist that identifies the most important age-inducing types of damage. His list has seven key categories.

First we would need to identify and repair two types of change that occur at the cellular level. One is that some cells have a tendency to die at a faster rate than they are replaced, leading to a depletion of particular cell types. In this case we would need to stimulate fresh growth. The other concern is the opposite issue – the accumulation of cells that are supposed to be dying and are not. Tackling this sort of cancerous growth is already the subject of considerable research.

Then the annual check would need to look for two different sorts of 'damage' that occur outside the cell. One is the accumulation of garbage that the body's mechanisms cannot digest and remove. These are normally large protein-based molecules such as the amyloid material that stacks up in various tissues, the most famous being the plaques that form in the brain in Alzheimer's disease. The other is the stiffening, the loss of elasticity of various long-lived tissues, like the lens of the eye or the artery wall. This stiffening occurs as a result of chemical reactions that occur in long-lived protein.

Finally we will need to tackle three things that go on inside cells. One is the accumulation of indigestible molecules of various sorts that can build up to the point that they physically prevent normal life inside the cell. Trying to make the cells work is a bit like trying to live in a hoarder's hovel, where discarded broken furniture and worn-out clothes occupy almost all the space. Given that this is going on inside the cell, it is a much harder problem than clearing the extra-cellular junk, but De Grey believes he has potential solutions. And the other two are both forms of genetic mutation. One is a mutation of the genes in the chromosomes stored in the nucleus, the sort of thing that causes cancer. The other involves mutations that occur in a special part of the cell called the mitochondrion, which has its own DNA. It's the only part of the cell that has its own DNA, apart from the nucleus.

A mutation occurs when the code sequence stored in the genes inside a cell changes. As little as one 'letter' change in a sequence of thousands of letters can lead to the cell changing the way it works. Occasionally a mutation can improve a cell's function, but most of these changes will reduce the cell's ability to function.

Put this way, the list is quite short and concise. At first sight it might even seem relatively straightforward, but each of these seven items is a massive step in itself. It also strikes me as being dangerously close to a somewhat obvious wish-list of wonderful but unobtainable features. Also, a key part of this strategy is the realisation that there is no point in doing six of the seven features. Solving most types of damage is not enough, because the one you miss will be the one that gets you.

Solving most types of damage is **not enough**, because the **one** you **miss** will be the one that gets you

'Of course for each of these things, the devil is in the detail', says De Grey. 'We have to work out the absolute details of how to go about targeting this damage and removing it, or in some cases making it harmless, without having serious side effects elsewhere. And most of the things that we have to do come down to stem cell therapies, gene therapies, perhaps some tissue engineering of various sorts. Some of them are more traditional types of things, like vaccinations and small molecule, pharmacological approaches. It's a multi-faceted problem and it will be dealt with by a multi-faceted, multi-part solution, just as the repair and maintenance of any man-made machine.'

I'd love to get really excited about it, but my hypeometer starts buzzing. The adult human body has one hundred million million cells – give or take a few. If this approach is going to work, surely we are going to have to monitor each and every one of them, and repair the ones that are failing to deliver.

'It's important not to overstate that problem', De Grey replies when I put the issue to him. He points out that the reason we live as long as we do at the moment is because we have such amazingly sophisticated automatic in-built repair machinery. All that is needed is a small tweak to make up for a few places where it has not developed enough capability. He is looking to mildly augment the in-built natural repair machinery that's already encoded in our genes.

Longevity Escape Velocity

De Grey's second basis for his confidence is that he doesn't believe you will need to hit all the cells – he hopes that most of the cells will be enough. The reason for this is that his plan is to clear up the errors that you can solve now, and buy some time for science to move ahead. When the next lot of wear and tear begins to slow you down in a decade or two, you will be able to perform all of the previous fixes, plus the newly developed set. Even though the problem has got harder because the things that the therapies could not repair the first time round are more entrenched, he hopes that will be outweighed by the improvement in the therapies themselves.

It's what he calls the longevity escape velocity. I like the idea. A rolling program of maintenance that, like painting a bridge, means that once you get to one end you turn around and start painting it again. Each time you start the bridge is older, but the technological make-up of the paint is more able to support and protect. At the same time I wonder if it will work against all of his seven points. How about the need to check unwanted cell growth? If your maintenance program misses just one cell, you are left at the risk of that cell breaking free from normal control and triggering a tumour.

De Grey counters this by saying that if you protect 99% (I'm still not sure how, but let's put that to one side), then you reduce the probability of having a rough cell run riot. The problem here is working how much reduction you will achieve. Any simple calculation would depend on all cells having equal chances of kicking off cancer, and that isn't the case. If you miss any of the highly prone cells, then the reduction would be almost unnoticeable.

Potential Solutions

All the same, there seem to be many other parts of his plan that would work in this system of cyclic repair. So, I asked, are there any examples of practical things that you could do now?

Here we hit a problem. The human body, as we have already mentioned, does a remarkable job of maintaining itself, so improving on the situation is not straight-forward. Yes, therapies can distinctly increase the lifespan of people with particular diseases, so that if you have type-2 diabetes, you no longer die as soon as symptoms start to appear. But that is very different from developing strategies for extending normal lives.

Also there is that need to hit all seven features at once if you are going to see any life-extension. That said, there is progress in some of the areas. Tackling the accumulation of extra-cellular garbage seems to be moving forward with possible vaccination approaches that could give a person a built-in ability to prevent formation of the so-called senile plaques that plague people in Alzheimer's disease.

Stem Cell Solutions

Stem cell therapies are also showing distinct promise, particularly when the cells are taken from the patient. For example, there are many trials in people who have had heart attacks, and have patches of damaged muscle in their heart. To repair the damage, doctors take cells from the person's own bone marrow and inject them into the damaged patches. These cells repopulate the tissue and significantly reduce symptoms – this would be one way of solving the issue of depletion of needy cells.

Scientists are also gaining a much greater ability to control the way that muscles grow, and to understand how cells work together to form tissues like bone, cartilage and muscle. This will lead to an era of what some people call regenerative engineering, and the greatest progress is likely to be seen when we start to combine advanced material science and cell biology. Engineers are building scaffolds that can then be populated with cells, and thus create engineered tissues.

In addition, trials using stem cells have also been carried out in people with Parkinson's disease. Here the results have been mixed. Although some people saw benefits, others developed severe uncontrollable shaking, and as a result human trials are on hold.

One of De Grey's more ambitious plans is to work out ways of clearing debris from inside cells. When the cell doesn't want particular molecules, it sticks markers to it and sends them to a membrane-bound bubble within that cell called the lysosome. The inside of a lysosome is acidic and packed with enzymes that set to work breaking up incoming molecules. We already know of over 60 enzymes, each targeted at different molecules, and De Grey's plan is to pack more in. Drawing from work done in bioremediation of contaminated land, he is looking for enzymes that naturally occur in bacteria and fungi found in contaminated ground. Now if you want to find organisms that have adapted to break down the waste found in human cells, then how about heading to a graveyard? As the bodies decay, all of the molecules in all of the cells get munched up and recycled. De Grey's hope is that scientists will find the organisms that break down the contents of lysosomes, isolate the enzymes and then build them into gene therapy-type packages that deliver these enzymes to the lysosomes in living cells.

Never Too Late to Start

There is no doubting De Grey's enthusiasm. But, I wondered, how long would it be until we saw any tangible results? Could I be expecting to draw my pension for longer than I'd thought?

'Of course timescales in this are extremely speculative. We're talking about a highly ambitious technology and I think at this point, we have a 50/50 chance of getting there within 25 or 30 years, subject to good funding. But anything that is that far off, 25 years or so, is necessarily very speculative. If we get unlucky and we get problems that we haven't yet noticed, then it would be a hundred years', he replied.

For De Grey, the point is not so much about the date when we arrive at a working solution, but the need to start heading for it now. 'My point is it doesn't really matter how long it's going to take, for two reasons. First of all, a 50% chance of success is quite enough to go for, quite enough to motivate . . . , and secondly, if it takes longer, the sooner we start, the sooner we finish. We'll just be saving a different set of people's lives', he continued.

I'm intrigued by how high a figure of probability that De Grey is claiming – a 50% chance of success in only 25 years. Given the speculation that he's already mentioned, this seems remarkably optimistic. Would it be worth pursuing on a 1% chance? I can feel my 'hypeoscope' beginning to buzz . . .

I point this out, but he is undaunted. 'This technology will arrive sometime, because any technology that people want arrives eventually, if it can . . . but if we get on with it now, it will arrive that much sooner and bringing it forward from a hundred years hence to 99 years from now still saves 30 million lives', he insists.

I don't think it is unduly cautious to point to the 'if' in that statement: 'if' it can. That, of course, leaves open the possibility that it can't. You only have to look at some 1960s comic books to see dreamed of technologies that seem further away now than they did then. I've yet to see anyone flying to work with a jet-pack on his or her back. The 'if' is an important caveat, but one that De Grey effectively dismissed. His reasoning for riding so quickly over the 'if' is that he really believes the scientists he and others are sponsoring will come up with the goods. Time will tell.

Is SENS Enhancement?

SENS is therefore a system that proposes to let our normal body continue going for longer. As such it doesn't plan to give us any new abilities, except for the ability to live longer. If it works, some of that capability will come from built-in modifications to our cells, and some from regular or periodic therapies that we receive. De Grey's aim is to address the shortcomings of the human condition, but not really to enhance what it means to be human. It's worth noting that this contrasts with some of the definitions of transhumanism that seek to let enhancement move us to a post-human position.

So where did this get us? De Grey is keen to give us the opportunity to live a long life, if we choose. He becomes less easy when it comes to saying that long life is an enhancement, and as a consequence is keen to give us easy ways to end our lives when we make that choice as well. If it worked, that would probably become more important as one of the recurring fears of anti-ageing is that we will be forced to endure a long, low-quality life. If you had taken away the natural means of ending it, you may need to build in some active steps.

But where does it fit on the hype–hope scale? The hope is certainly high, but much of the technology De Grey alludes to is far from being ready to implement. This form of enhancement is great fun to muse about, but the requirement to get the whole package right before you see any real benefits means that I'm not holding my breath in anticipation of any quick fixes.

The Mighty Mouse

Look after a laboratory mouse and it will live for between 500 and 1000 days – give or take. Feed it carefully and more importantly manipulate a few of its genes, and scientists are managing to stretch that figure.

De Grey is keen to encourage this, and so has established the Methuselah Foundation's Mprize. This rewards any scientist who manages to make his or her mice live longer than any previous mouse.

The current title holder, Andrzej Bartke from Southern Illinois University School of Medicine, USA, claimed his reward in 2003 for a genetically manipulated mouse that lived 1819 days (http://www.mprize. org/).

Laboratory mice can be a great test bed for new ideas and potential enhancements.

©ISTOCKPHOTO.COM/DRA_SCHWARTZ

Alzheimer Vaccine

A key element of Alzheimer's disease is the build-up of small proteins called amyloid beta (Ab) peptides in the brain. They build up as plaques and these plaques are accompanied by memory loss and mental deterioration.

One possible approach to preventing this material accumulating is to inject bits of the Ab protein into muscles. The body's immune system recognises this as 'foreign' and builds antibodies to destroy it.

When Japanese researchers did this using experimental mice that are genetically engineered to get Alzheimer's-like symptoms, they found that the antibodies crossed into the brain and significantly reduced the amount of Ab plaques compared with untreated mice.

It appears the vaccine was indeed protecting the mice, but will this work in humans? The initial trials in 2001/2002 were discontinued after at least 15 patients of 80 patients developed inflammation in the brain. Pharmaceutical giants Elan and Wyeth then started new trials in 2006, using a vaccine created by injecting the amyloid material into rabbits, and injecting the antibodies they produce into patients. They hope a large clinical trial will be under way by 2008.

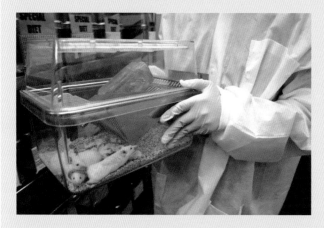

3 Uploading

In my quest for an enhanced lifespan, I'd looked to technology to replace vital bits as they cease to function, and found that while cyborgs are alive and kicking, they have a long way before they can deliver something that looks like a clear enhancement. I'd also discovered that we seem to be a long way from radically extending the number of years that we can live with our cells and the bodies they build. So I'm left with the option of climbing out of the body, leaving this old biological frame behind and starting a new existence in a new state of being.

One of the more radical ideas that is bandied around is uploading – the concept that you suck the bit that is 'you' into a computer and let it live there. In an age when we celebrate our ability to think, this 'you' is often reduced to your mind. The idea is not so much to become a dis-embodied self, but to become a re-embodied being; to swap your carbon-based life for one based more on silicone. You would exchange an existence that emerges from biological cells for one held together by technological chips.

Uploading – the concept **that** you suck the bit that is 'you' into a **computer** and let it live there

In brief, the post-human dream is to find some way of interrogating a person's brain to discover in minute detail how it ticks and what memories it holds, then you can capture every bias in opinion and effectively catalogue every mood. This could be transferred onto some yet to be developed supercomputer, in which the person could live.

This digital existence comes with a number of new possibilities. There is presumably no end to the number of days, decades or millennia over which your new life could span. As the set of processor chips and disc drives you inhabit ages, you could simply be shot through the latest equivalent of a USB port to a new machine. It could be like moving house. Maybe you would move to a similar machine; maybe you would upgrade. Each move to a new machine could offer a fresh start possibly for expanding your range of facilities. Climbing up the computer-based artificial estate ladder could enable you to do the equivalent of moving from a two-room flat, to a four-bed detached with conservatory and hot tub. Or you might decide to down-size, to slow down from the equivalent of a quad processor-driven high-speed chase, to a more sedate lifestyle in a single-processor cottage in the countryside.

Forming relationships with others would bring a new level to the concept of networking. It would be a matter of linking to the web and seeing who is out there, without knowing where they are. We are already used to the idea of downloading information and images from the web with no knowledge and few concerns about the location of the server on which they are ultimately held.

Enthusiasts suggest that interactions could be controlled with protocols – laws if you like. You would have to abide by the rules of your chosen cyber-society, but you could probably change societies if you wanted.

Backing Up

You could also create back-ups of yourself to protect against viral attack or sudden catastrophic failure of your main drive. So long as you back-up regularly the worst that could happen would be a week or so loss of life – a brief amnesic blip in the electronic stream of your being.

But would you start to get paranoid? How would you be sure that your back-up is safe from someone tampering with it? Who has access to it, and how do you arrange secure non-hackable systems to prevent others from finding your deepest thoughts against your will? A breach of security would be cyberworld's equivalent of forced entry – personal trespass. Hacking could be on par with rape.

Backing-up before key moments in life would enable you to have a second bite at the cherry. You could jump back to the beginning of the day and try it again if life's offerings were less fulfilling than you had wished. You would, however, need some way of remembering that you'd decided to try again, otherwise you may never get out of the 'go back to yesterday loop'. But if you did constantly return to a happier version of your life, would it really matter? Would anyone else actually know? If they did know, would they care?

This reveals one of the key problems with this extreme level of individualism – it only works so long as you don't interact with another uploaded being. As soon as you do, you and your electronic actions will be part of that other person's memory and cyber experience. If you reboot the day, will you require them to reboot as well, or will you have to put up with them knowing that you are having a second go?

Although an uploaded person may theoretically go back and start again, the rest of cyberworld will have moved on. You might have taken a technological attempt at thwarting ageing, but you haven't stopped the clocks of cyberspace.

How about creating multiple copies of your identity? Just think, rather than repeating days as though you were the only living creature and everyone else was a stooge, what if you were tempted to put yourself in more than one place at the same time? Each copy of yourself would diverge from the original. Imagine the rivalry if the two of you met!

In the 1993 film *Groundhog Day*, Bill Murray plays a self-centred and cynical TV weatherman who is sent for the fifth time to the small town of Punxsutawney, Pennsylvania, to cover the groundhog ceremony held every 2 February. He stumbles into a time warp and lands up repeating the day time and again. Murray benefits from this, not because he starts again each day with no prior knowledge of living it before, but because the rest of the town does. Each day he learns more, and consequently becomes more capable of manipulating events. He doesn't reboot himself, in effect he reboots them. The whole plot would fail if anyone else had the ability to play the day again.

The Matter of Mind

Our speculations seem to be getting silly. I still needed to be convinced if there was any real potential in these ideas. It is all very well for someone to speculate about possible technologies, but I wanted to find out what was genuinely likely and what snags stood in the way and would need overcoming if a mind-machine was going to appear.

I headed to Oxford, and near to the historic castle I found Anders Sandberg, perched at his desk at one side of a large open-plan office and meeting room at the Oxford Uehiro Centre for Practical Ethics. Sandberg is a Swedish, fair-haired, gentle giant, who moves carefully and speaks softly, but is bubbling with ideas and constantly commenting on the technology in the world around him. Just walking up with him to his first floor office had prompted various mini lectures on the technology of the automatically opening door, the lift and the coffee maker.

Sandberg is enthusiastic about the possibilities of creating thinking computers that can emulate human intelligence and has just finished writing a paper that reviewed the work already done, and outlined what is still unknown. He seemed an ideal person to ask how you deal with the problems associated with uploading and he opened our discussion by suggesting that there were two classes of problems that need to be overcome with uploading – technical ones and philosophical ones.

We started by discussing the philosophical issues, largely because these were the ones that he described as flypaper – if you are not careful you simply get stuck on them and get no further. The reason for this is that the issues are too big to solve and include questions such as, What is it that is actually 'you'? Is it just your mind, your memories and your modes of thought, or is it more? To what extent are you shaped by your body, or are you some form of homunculus, a sort of spiritual being sitting inside a machine? And then there's the question of the balance between your physical make-up and your experience – the result of the body's interaction with the brain, and of the brain's interaction with the external world, as mediated through the body.

There is nothing new to realising that this is a difficult area of thought and has been chased around for centuries and millennia. For example, the idea of having a material, physical body and disembodied thoughts is a concept much loved by dualist philosophers such as the seventeenth century philosopher Descartes, but criticized by many since.

In the seventeenth century, French philosopher René Descartes proposed that people were composed of two sorts of stuff – spiritual stuff that made up the mind, living inside a body made of physical stuff. It was an issue of mind and matter, with mind being critically important – mind over matter.

This fits with the idea of a humunculus, a being that effectively inhabits your brain. In this model, the humunculus is the real *you* in the biological machine.

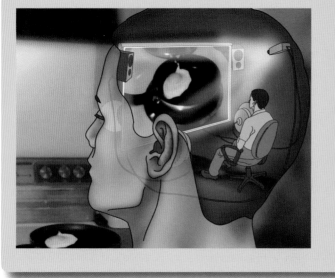

REPRINTED FROM HTTP://EN.WIKIPEDIA.ORG/WIKI/IMAGE:CARTESIAN_THEATER.JPG

Intriguingly, Descartes proposed that the mind interacted with the body via the pineal gland. In some ways this proposed mechanism is not a million miles away from how the hormone system does actually work.

Enhancement enthusiasts argue that you are a single total being, that your mind is shaped by the machine it inhabits. The classic example is railway worker

Phineas Gage, who, on 13 September 1848, had the unfortunate experience of having a 1-metre-long, 1.3-centimetre-wide, steel rod blasted through the front part of his brain. Remarkably he was only unconscious for a few minutes and after six or so months he went back to work, but his personality was permanently altered. Gone was the affable Gage and in his place was a temperamental and antisocial character – his friend said he was no longer Gage.

Since then, medical history shows that people who have brain tumours that put pressure on the frontal lobes in the brain often suffer significant changes in character. Coincidentally, brain scanning shows abnormal activity in this part of the brain for some people who have difficulty forming social relationships. The

Medical imaging wasn't invented when Phineas Gage had his accident, but this computer-generated graphic shows what a scan could have revealed.

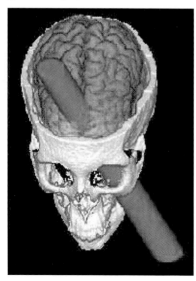
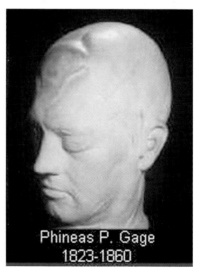

fabric of a person's brain is therefore an important element in creating who they are, and the physical stuff of your existence should not be easily dismissed. So is it ever conceivable for a conscious person to remain the same once he or she has had such a radical shift in the nature of being and the consequent change in the inputs and outputs of life?

The issue is not easily resolved, as there is little real understanding about what actually constitutes consciousness, what makes the mind. If you can't determine that, it is going to be difficult to determine that a machine either does or doesn't have it.

'What would you have to do to still be a mind? And how would you know if a computer was conscious?' I asked Sandberg, not sure whether it was a profound or stupid question.

'I think the best thing would be to try it and have a philosophical debate with an uploaded mind about whether it is a mind or not, because I don't think that philosophy can actually solve the issues of personal identity', he replied. 'Maybe we are going to get some headway about what constitutes consciousness before we get to uploading, but it's going to be terribly tricky.'

Computer Solutions

When we turned to the practical aspects, Sandberg became more enthusiastic. With a background as a computational neuroscientist, he has experience of an academic world that is already simulating simplified neurones. The simplified neurones have the advantage of being easier to run on computers than complete neurones, and don't need a supercomputer to run them. Realistic neurones are of course much more exciting to run, but then have the disadvantage of requiring a supercomputer to drive a handful of simulated neurones.

The information storing and processing part of the brain is made of spider-like cells called neurones. They are spider-like in that they have a central body and spindly legs; the difference is that the legs branch repeatedly until they create up to 10,000 endings. These endings reach out and make contact with other parts of the same neurone and with other neurones. The touching point is called a synapse and is the place where signals are transferred.

At the top of the simulating tree, computer scientists have got to the point where, with the aid of the world's largest available computer, IBM's Blue Gene, they can simulate a network of 22 million neurones connected in a web of half a billion links. Before you get too excited, running the machine for a few hours managed to simulate one second's worth of brain activity corresponding to a few square millimetres of the surface of the brain, or the cortex.[1]

On the positive side, the computer showed that massively interconnected systems of neurones could start to develop organised behaviour. When one part of the network was stimulated, another rose from its ground state to an active memory state within 50 milliseconds. The scientists believe this 'emergent behaviour' is one sign that their machines are beginning to mimic small patches of a working brain – to mimic life.

Time for a reality check, though. Right now we still have a very basic understanding of how the brain works, so are not clear what to look for to see whether we are

[1] Project report for Blue Gene Watson Consortium Days: massively parallel simulation of brain-scale neuronal network models. Mikael Djurfeldt, Mikael Lundqvist, Christopher Johansson, Örjan Ekeberg, Martin Rehn and Anders Lansner. KTH – School of Computer Science and Communication and Stockholm University, 18 October 2006.

mimicking life. That doesn't mean there is no value to experiments. They are fantastic, but they have more power in trying to establish the nature of normal brain function than they do in providing a future home for wandering human minds. In addition, while the feat is impressive, we need to realise that each simulated neurone only made an average of 500 connections, rather than the 8–10,000 or so connections made by neurones in real brain tissue, so it was still significantly less sophisticated than real tissue.

'Building a 22 million neurone network was of course a technical tour de force, but it was not really an image of a real brain. It was just neurones organised in roughly the manner we think they're organised in the brain – it's interesting on its own', explained Sandberg. Undaunted by this admission of the real achievement, he then went on to claim that some of the simulations are so sophisticated that biological scientists cannot tell the difference between the computer simulations and the real thing. 'That doesn't prove anything really, but it's a very crude first start', he said. To write a big computer program that manages to mimic the action of a network of cells is very different from saying we have made a machine work in the way that cells work.

Even if some computers are beginning to act in the way that biological brains act, to my mind, however, there is still a huge leap from using the world's largest supercomputer to simulate a few seconds of operation of a small patch of simplified neurones, to a claim that we can successfully upload a mind. Sandberg, however, seemed undaunted by the gulf he needs to traverse in his leap of scientific faith.

The Problem of Place

Knowing where you are in a familiar room with your eyes shut comes down to place cells. These neurones get excited when you recognise where you are in a room, and they then talk to each other so that you can navigate your way around.

Researchers study these cells by placing a mouse in a cage and monitoring some of these cells at work. They see different ones light up as the mouse moves to different parts of the cage; the nearer the mouse gets to the place cells' location, the more vigorously the cell fires. The mouse therefore recognises where it is by assessing the strength of activity in many thousands of these cells.

At first thought, you could suggest that determining the pattern of connections among these cells would let you map the mouse's internal world. Sadly, that wouldn't work. Mice seem to use the same cells to map their places in many different locations.

Knowing the network map alone will not bring the mental map to life. Something much cleverer is going on.

Looking Ahead to Today

I guess my scepticism was enhanced when I read an academic paper by Nick Bostrom entitled, 'How long before superintelligence?' Published initially in 1997, this started by defining superintelligence as 'an intellect that is much smarter than the best human brains in practically every field, including scientific creativity, general wisdom and social skills'. Quite a challenge, then. He points to some of the potential problems involved in creating a computer that would be powerful enough, including costs. Back in 1997 it took about 400 engineers to build, and a new chip factory costs over $2 billion to build, and there is nothing to suggest that things have got any cheaper.

He predicted that a computer capable of emulating the human brain would be available by 2004 and that Moore's law would run into the buffers in 2007. Neither of these has occurred. In fact he has a section in which he discusses why previous failures of artificial intelligence should not be taken as an argument against future success. A 2005 postscript added to a web version of the paper (www.nickbostrom. com/superintelligence.html), however, is less upbeat, and includes the statement that 'No dramatic breakthrough in general artificial intelligence seems to have occurred in recent years'. He also states that previous estimates of the computing power required to emulate a human brain severely underestimated the scale of the task.

I mention this not to poke fun at what people have written in the past, but just to show how the enthusiasts for the technology have a track record of making statements that can't be fulfilled.

Slice and Scan

The next technological issue is how to get the information out of someone's head to load into your multi-mega-network. 'So the first thing you need is a scanner', Sandberg explained. 'You need something that can scan a brain at the right resolution.' He made it sound simple, explaining that it is quite routine in biological labs around the world to take small chunks of brain, about 1 cubic millimetre or so, treat them, freeze them and slice them into micron-thick slices. You can now scan them with an electron microscope and analyse the images.

Moore's Law

Way back in 1965, one of the co-founders of the computer chip manufacturer Intel, Gordon E. Moore, made a far-sighted speculation. He said that, for a fixed price, the number of active elements in a computer would double every two years. He thought that this would probably hold for about 10 years.

In fact, history has shown that 'Moore's law' has held ever since. Every two years computers get twice as powerful.

The question now is whether this can continue forever or whether we will reach a point of miniaturisation where we can go no further.

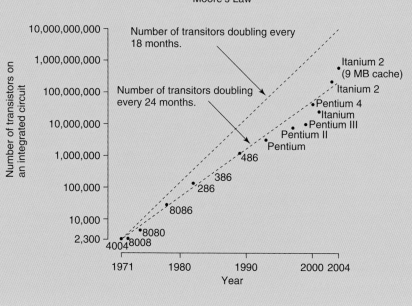

Moore's Law

❚❚ You need something that can scan a **brain** at the **right** resolution ❚❚

But then you hit two issues. Firstly, there is the sheer scale of multiplying this to take on a whole human brain, and secondly, if you did what would you look for? Oh, and thirdly the person would need to be dead before you could start.

If we put the problems of getting the brain out of its head to one side, we still need to consider the complexity of the system. At its peak, a human brain has around one trillion (10^{12}) neurones. If each of these has 10,000 synapses, this gives the brain 10 quadrillion (10^{16}) possible connections. As you age you lose some of these so that by adulthood you probably have a mere 1 quadrillion. Just counting them would be vastly more than any computer can achieve now or in the near future.

The exact figures are up for grabs. Sandberg thinks that a trillion neurones are possibly 10 times too many, but that would still leave one quadrillion connections. It's a huge number, but that is just the sort of thing that computers are good at coping with, indeed some recent graphics cards can actually do a teraflop, one trillion (10^{12}), operations per second. That would allow them to count to a quadrillion in just three hours. Part of the speed is that they act in parallel, which allows one part of the chip to count one thing while another counts another. If you are planning to try and upload someone, this is precisely the sort of technology that you would look to, and you would probably ease the strain by sharing the workload among a lot of computers.

Counting connections is the easy bit, though. To make sense of a person you would need to do much, much more. To start with, you would need to determine the connection's relationship relative to each other. From what we know about the way our brains work, part of the process of recording memories seems to be achieved by the neurones altering the arrangement of these connections. The exact pattern, and alterations in patterns, seems to be critical.

Even if you could **flash freeze** a **living** brain, you would **grab** a **single moment** in time; **not** its dynamic **nature,** with the connections constantly changing in a **frantic** dance

This points to another issue. Even if you could flash freeze a living brain, you would grab a single moment in time, not its dynamic nature, with the connections constantly changing in a frantic dance. You would have a snapshot, when instead life is more like a movie. We also don't know whether it is the static positioning of neurone synapses or their dynamic movement that creates and holds memories and processes ideas. If the movement is important, then uploading is in even deeper trouble.

The point where two neurones meet is complex and under many different forms of control. It is proving a challenge to understand individual synapses – much less billions.

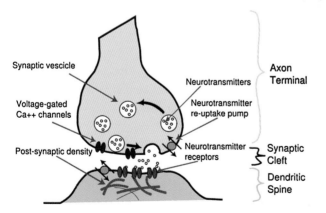

Synaptic vescicle

Neurotransmitters

Axon
Terminal

Voltage-gated
Ca++ channels

Neurotransmitter
re-uptake pump

Post-synaptic density

Neurotransmitter
receptors

Synaptic
Cleft

Dendritic
Spine

It doesn't stop there. Just knowing that there is a connection between two neurones – a node in the network – is interesting, but to know what it is doing you need to analyse the chemicals in the junction. Messages are passed from one neurone to another not just by touch, but by passing and receiving neuro-transmitters. To know what is happening at a synapse you need to know not only which neurones are connected together, but how much of each neurotransmitter is passing between them, and also how many receptors are present on the receiving cell.

On top of this neurotransmitters are removed from the synapse so that they don't just act forever. Drugs like Prozac affect this process. Knowing the way that these reuptake mechanisms are tuned will be critical to knowing what is going on at each terminal. Without all of this the information may be interesting, and may shed light on how a brain works, but just having the information will not let anyone record 'you' on disc.

Although Sandberg is well aware that any attempt at uploading will require a chemical image, he remains buoyant on the basis that he has a good understanding of the problem, even though no one yet has a solution. There are some interesting approaches that might help. Immune staining techniques where you can stain for different neurotransmitters could help, and fluorescence patterns can squeeze a lot of information out of biological tissue. But to make it work you will need to build a weird hybrid scanning system. 'It can't be a traditional optical microscope or electron microscope', he concludes. 'It's going to be some combination of several techniques; it's going to be a new kind of microscope that's not yet built. The pieces seem to be around, more or less.'

An alternative approach is to see whether you could infer the type of neurone just by looking at its shape. If this were possible, it would be a great short cut. According to Sandberg, though, half of the scientists who have thought about it are pessimistic and don't think this short cut will work.

To inject a slice of reality into my thinking I emailed Professor Peter Peters, a Dutch scientist who works at the forefront of electron microscopy and passionately pushes this mode of imaging to the limits. He pointed out that you can currently only scan about 0.2 cubic millimetre of brain tissue per day. Given that the average human brain contains about one and a half litres of material, this would take in the order of 190 million days. To investigate 20 different possible receptors at each junction in the network currently takes on the order of one year per junction using cryo-electron tomography and cryo-immunogold electron microscopy, and as we said there are billions of junctions in there. 'But', he added, 'these things might be totally different when we retire.'

'It is very clear that we cannot use just one microscope for the scanning. My own vision', says Sandberg, 'is a kind of uploading factory, with hundreds or even thousands of microscopes doing automated scanning at the same time. The logistics is pretty mind-boggling.'

Even if this could be built, the idea of scanning is so that we can replicate the network, and I still wonder whether we are in danger of assuming that the brain works like a computer, because we can make computers and know how they work. In reality, the neurones of the brain are vastly more complex than a network of connections. The timings of each of the myriad of signals that arrive at each point in the network are as influential as the shape of the network itself. Knowing the shape is just the start.

The Living Dead

This sort of scan technique would obviously destroy the original. You would need to take someone before they had died and salami-slice their brain. It is an issue that troubles Sandberg and he's been trying to find a non-destructive way. To get all the information from the brain without slicing it means that you have to work from outside through several centimetres of tissue. While doing this you need to achieve a resolution down to a few microns. That puts a horrible constraint on any technique, particularly as a living brain is moving around. Sandberg is the first to admit that there doesn't seem to be any technology that can do it.

One possible option for non-destructive data acquisition could come from the world of nanotechnology and micro machines. Sandberg believes that in principle you might flood the brain with nanomachines and let each nanomachine plug into each neurone. The nanomachine's task would be to figure out what that neurone was doing and feed the information back through a wireless or optical network to the outside where it's gathered up. 'I think it might be physically possible in the sense that such a structure can exist, it's just that it might be terribly hard to actually build!' admitted Sandberg, adding that Robert Freitas, one of the gurus of molecular nanotechnology, believes that it could be possible one day, but it's going to be a terribly involved technology, highly complex. 'You move into the realm of hypothetical nanotechnology', comments Sandberg, 'it's still very unconstrained so far, and that annoys me; I like being a little bit constrained by reality.'

Forget Uploading – Try Emulation

Sandberg told me that the term 'uploading' evokes so much science fiction that most scientists try to stay away from the word. I'm beginning to see why.

Undaunted, he continued. 'The most likely scenario right now would be that you take a brain, you freeze it into liquid nitrogen and then you slice it carefully and scan with an electro-microscope. Then you put the information into a computer and do a lot of image processing to figure out the shapes of the neurones, and which are connected to which – and of course you're going to need a computer able to run all this.'

Even by current standards of imagination, I think that such a computer will be almost unimaginably powerful. But is that just where my imagination is too limited? There is, after all, more computing power in the average mobile phone than the entire Apollo space mission. The decades since the 1969 moon landing have created an explosion in our ability to pack a punch on a chip, and the rate of expansion still hasn't shown any sign of peaking.

If Moore's law holds, Sandberg believes that you could reach the scale of information storage and processing you need by about 2030–2040, and somewhere around 2040 that enormous computer might even fit on a desk. The problem is of course that even if Moore's law does follow a perfectly straight line, it might break at some point. Like many pundits, Sandberg reckons that Moore's law will hold for the next few years, but that no one could say with confidence how things will develop over the next three decades.

It's quite a large extrapolation to assume that growth will continue to 2040. Not many people have made accurate predictions of where technology will be over three or more decades.

Green Living?

Sandberg also raises the potential environmental benefits of an uploaded life. No need to commute to work, burn fossil fuel to fly to some exotic vacation location, or chop down rain forests to build garden furniture. It's a concept that is open to critique – the sort of computer that is capable of such a high level of performance has not yet been developed, and ones that are millions of times less capable do not fit into your pocket. They still inhabit sophisticated cabinets in air-conditioned rooms with energy needs. The hardware is made of rare materials that are environmentally costly to mine and refine, and not easy to break down and recycle. And unless someone builds self-building machines, presumably you will still rely on someone, some human, to maintain your considerable power supply and requirement for new kit.

'I think you are underestimating just how efficient ultra-advanced computers could be, but this is of course conjectural. It certainly doesn't seem likely that uploading could ever happen before available computer power becomes very dense and cheap – if it takes a major mainframe to run each upload, very few uploads will be made. Even so, with a fairly big computer it is quite possible that it would be greener than our current lifestyle. Consider the resources it takes to produce the 500 kilograms of food we eat every year', commented Sandberg.

Where's the New Me?

A final and rather critical issue to ask is what you would have at the end of the process. If, for example, you allowed your brain, before you died, to be removed, frozen, sliced, scanned and stored in a database, what would you be when the machine was booted? This is moving us away from the practical issue, back to Sandberg's philosophical flypaper.

Here I was slightly unnerved by the lack of certainty in Sandberg's answers. If I want to upload myself, then it is rather important that 'I' still exist at the end of the process.

'Whether that system will now experience consciousness – be the original person – that's an interesting question', mused Sandberg. 'I think it could if everything is done properly and we actually got all the science at that point right – it will work. But of course we will have to try it to find out.'

One problem is that even if the computer is programmed to say 'Hello, I'm Pete Moore', how would I know whether it is a fake, an impostor, a cartoon representation that bears only a passing reality with my former self?

But Sandberg dismisses the issue quickly, saying that personal identities are not clear things, but rather messy and soft, with blurred borders. Above all, he says, they are not very stable and certainly change as we grow older. 'So if we can handle growing older we can probably handle being translated to a computer', said Sandberg, in what again seems to be a giant leap of philosophical thought.

It seems to depend very much on your definition of identity. 'It might be that for people with different definitions of identity, some of them might survive being uploaded and some might not', he muses. 'And even if uploading doesn't produce an identity it's still a very interesting process if it just creates a mind or a mind-like thing.'

One consequence of this is that Sandberg prefers to use the term 'whole brain emulation' to 'uploading'. It is an important shift. Gone is the idea that you take a person and move them into a new realm. In its place is the attempt to build something that emulates a brain. Now there is no drive to bring an existing person back in a digital form, but to create a new being based on a more generalised human template. Is this really human? I think not. Is it post-human? Again I think not. What you have is a technology that wasn't capable of copying a human, and so went off in a different direction. Is it likely to be highly capable? Yes, most certainly, but surely that is different.

If you were looking at uploading as a method of extending your life indefinitely, of enhancing your existence, this subtle shift seems to serve a critical end to your hopes. If on the other hand you wanted to build something that used humanity as a starting point, and then headed off on a technological adventure, then this could be the first critical step.

I'm still left wondering. How would you know whether what you had created was really a human in a box? For Sandberg, the golden test is still the dear old Turing test. This was first proposed in 1950 by Professor Alan Turing,[2] and says that a machine has passed the test if a human judge cannot tell whether she or he is having a conversation with a person or a machine. The problem with this is that you will always be left wondering if there was still one more question that could reveal a distinction. It could never give a conclusive answer.

However, there is distinct value to a mechanical mind even if it is not human. We have seen computers enter many fields of the workplace, and if their power started to stretch to the point that they could be said to have 'minds', then employment prospects for many people would change. A mind that we could copy at will and still did not ask for holidays, pay rises or sympathetic leave would be welcomed in many institutions.

Perhaps Descartes could show us a way forward with his famous statement *Cogito ergo sum* – I think, therefore I am. At the root of this is the idea that the only thing we can rely on is our ability to think. All else could be an illusion. So if the computer can 'think', then 'it is'. The problem, though, is that only the computer would know. You cannot use this formula to ask whether anyone else exists. I think, therefore I know that I exist, but you may still be an illusion, or just a figment of my imagination.

[2] A.M. Turing, Computing machinery and intelligence. *A Quarterly Review of Psychology and Philosophy.* 1950; 59(236): 433–460.

Uploaded 'Animal' Rights

Undaunted by the near impossibility of knowing whether an upload – an emulated brain – is really a mind or a fast machine, upload enthusiasts are keen to discuss the ethical status of their proposed creations.

What do you do, they ask, if once we have made the first upload, once we have got a mind in a machine, we might not get it quite right and build a mind with a hangover? Then they ask whether it would be ethical to tinker with the machine to try and dry the being out. Would you be allowed to mess with something that is liveable, but living at a low level? Could you change the mind on a deep level?

And at what level do you start to consider the need to invoke an Act of Uploaded Rights? If a system can only emulate a snail, then few people would worry. How about a mouse? If you claim that an emulated mouse brain has the same mental state as a biological mouse, you would presumably be forbidden in the UK from experimenting on it in any way that invokes pain without some form of Home Office research licence. Sandberg raised the possibility of animal rights activists standing outside the door if I do that kind of experiment. If you worry about mouse-capable computers, then of course you will be much more worried about one that emulates a monkey.

To my mind a leap of logic has slipped in here. The crunch comes with the 'if we claim that an emulated mouse brain has the same mental states as a biological mouse . . .' because if we don't believe this, then there is no ethical issue. Personally, I would take a lot of convincing before I fell for that particular 'if'.

II Brighter Than Life

You don't need me to tell you that nestling inside each of our heads is a remarkable organ – the brain. Often described as having the consistency of cold porridge, it is fragile and prone to damage. Its survival is dependent on being encased in a thick globe of bone – the skull, and held firmly in a tight membrane rather like a shrink-wrapped cabbage.

At one level, this brain is a multi-tasking wonder. It enables us to make sense of our surroundings, integrating information about smells, sounds, textures, tastes and varying patterns of light. At the same time it stores past experiences, it processes data and creates new ideas. Alongside this, it manages the overall sense of well being, driving our desires to reproduce, tempering our moods, handling individual responses to music, art and dance. As if this isn't enough, it looks after all the stuff we seldom think about. Every breath we take is driven by nerve cells in the back of the brain; each heartbeat's timing is affected by neurones that respond to the body's need for oxygen. Movements of our gut, the cooperation of blood vessels and the timing of sleeping, waking and cat napping are all organised by the mush in our skulls.

I guess I was a bit surprised when Anders Sandberg complained that the human brain was really rather simple, that it could only do a very few things at once. What he was referring to, however, is just the little bit of activity that we do tend to think about when we consider the brain – our ability to think. And here Sandberg has a point. You can argue all day as to whether men or women are better at multi-tasking, but in reality few people can handle more than two or three tasks at once before they all start to grind to a halt, and in most cases what people are doing is not really handling them at once, but chopping each task into packets and juggling them so that while they handle only one activity in any given second, over the course of a few minutes or hours a number of different tasks have been accomplished.

This sort of serial processing, of taking tasks, splitting them into chunks and streaming them through a processor, is the stuff of old-fashioned computers. Sandberg would like to move to a situation more akin to the new range of parallel processors that have multiple sets of resources and handle multiple tasks simultaneously.

But once you take into account the housekeeping jobs that the brain is doing, the difference in capability seems less severe. Indeed you could argue the case that each of the millions of neurones in our brain is its own microprocessor and that we are constantly massively parallel processing.

The rate at which computers can operate is also staggeringly fast, with many millions of 'yes/no' decisions being made each second. Again, at first sight this is more impressive than most people's brains that sluggishly try to remember their seven times table, or pause in confused embarrassment in front of racks of bottles on a supermarket shelf trying to compare the costs of three competing offers on wine.

Sandberg is looking for a way of boosting the way that the bits of our brain that we take notice of each day operate. He would like to think faster, to handle his next book at the same time as thinking about dinner and planning his forthcoming holiday, giving each task totally undivided attention. He would like to stay alert for longer each day, to maintain concentration without becoming tired and consequently to increase his productivity.

He is far from being alone in this desire. We hear of college students on pills during exam time, or members of boardrooms loading themselves with caffeine to keep up with the pace. We carry ever more powerful pieces of information technology in our pockets that attempt to make up for our forgetfulness by giving us easy access to stored diaries, contact details and years of backed-up emails.

In the world of enhancement, the question is how far can we push these add-ons. To what extent can we integrate technology into our brains so that we can boost function? How much can we enhance our brains' ability to stay alert? And how much would you be prepared to risk and pay for these enhancements?

4 Let's Stretch

Drugs are one way of altering brain activity. Ecstasy, marijuana and all the naturally derived psychedelic drugs – mushrooms and things – give access to alternative perceptions of life and the world. No implants or extrasensory organs required, and many creative folk in studies, studios and on stage have used them to bolster links to the subconscious and to give them an altered focus and meaning in their work.

A Trip of Invention

Take Sam Patterson, the man who invented the Grip Shift®, gear changer for mountain bikes and the creative force behind SRAM Corporation, a big player in the mountain bike racing community. He puts much of his inspiration down to mind-expanding drugs, and some of the software writers who dreamed up far-reaching internet protocols and computer code believe that their power to think further came from use of mind-expanding drugs.

Psychedelic Prohibition

It's worth noting that psychedelics have very ancient roots. In many cultures throughout the last several thousand years of human history, psychedelics have been used – peyote cactus in North America, psilocybin mushrooms in Central America, *ayahuasca* in South America, *iboga* in Africa, *Amanita muscaria* mushrooms in Europe and ancient India, and so on. These drugs were used in these cultures as rites of passage, religious sacraments, or to treat ill patients in healing ceremonies. Viewed this way, our society is actually quite unique in that there are no socially sanctioned rituals involving psychedelics.

Our **society** is actually quite unique in that there are **no** socially sanctioned **rituals** involving **psychedelics**

Far from it. Although there is an active lobby for the legalisation of some of these drugs, particularly cannabis, and for their use in medical applications, they remain firmly illegal under the 1961 Single Convention on Narcotic Drugs, the 1971 Convention on Psychotropic Substances, and the 1988 United Nations Convention Against Illicit Traffic in Narcotic Drugs and Psychotropic Substances.

There is an argument that the safety problem is more to do with fitting in with society, than the effect of the drugs. If you were allowed to take it in a controlled setting, then its use would be much safer. I'm not sure, however, how you would set that sort of system up.

Legislation Links

Furthermore, I'm less than convinced by such a relaxed attitude towards psychotic drugs. The old adage that if you can remember the 1960s you weren't really there stands as more than a reliable quip. In reality psychedelics can have serious effects on people's short-term view of the world, and tend to have long-term consequences, not least of which is memory loss and unpleasant, sometimes terrifying, flashbacks.

The prohibition against use is not just a slap on the wrist to stop people having fun, but a serious attempt to prevent people harming themselves, and more critically to prevent them from harming each other. Jag Davies, editor of *MAPS*, a journal dedicated to collecting information about safe use of psychedelics, argues that it is the prohibition, the socially undesirable status of the drugs, that causes the danger and points out that they are not seen as dangerous in other cultures. However, one problem with the 'well they do it, so it must be okay', is that female genital mutilation and child labour are both central to other cultures but frowned on in our own, so I'm not convinced that the fact it is used elsewhere is on its own a water-tight argument.

Whether these protective goals are best served by restricting access or regulating use is outside the scope of this book. I want to step aside from the legal issues and concentrate on whether chemicals can enhance human beings, or at least enhance human experience.

Therapeutic Targets

Psychedelic drugs can have clear therapeutic effects. As well as the often-discussed pain-killing effects, there is evidence that the compounds can be useful in psychology and psychotherapy for certain conditions. They seem to work by breaking down the barriers with the subconscious.

In the 1950s and '60s, thousands of individuals were legally administered psychedelics in experimental treatments by mental health professionals, often in conjunction with conventional psychotherapies, to treat addiction, post-traumatic stress, end-of-life anxiety and many other intractable mental health problems. In addition, many normal, healthy people underwent psychedelic therapy to gain self-insight, stimulate creativity and to live happier, fuller lives.

There are anecdotal reports of a clear improvement to creativity, empathy, mood, endurance and other human functions. One probably un-resolvable question is whether many more people would report this if the substances they were using weren't illegal. And this points to one of the key difficulties in trying to work out what happens when you take mind-expanding drugs. The stories become anecdotal and often anonymous. It is difficult to get people to speak on record about actions that could land them behind bars.

Business

In capitalist economies, the push for success is certainly high. People in business spend shed loads of cash on training courses and mentoring programmes. They buy endless self-help books, each one promising to explain how to climb to the top of the money mountain and become a success by the time you are 40 and then how to have a fun-packed 60-year retirement spending the stashed-away wealth.

The issue is to get one step ahead of everyone else, to see the problems more clearly, to analyse more thoroughly, to assess more deeply; in essence to think better. But why go through all the hard work of scouring the shelves for the mind-expanding book to suit your needs – and there are plenty to choose from – when you could instead pop a pill and go straight to the front of the fortune queue?

Rumours abound about college kids illicitly taking Ritalin or similar drugs in an attempt to make the grade, but how many are really doing mind drugs is very difficult to pin down. One thing is certain, however: the pressure is there. Social re-shapers may want to create competition-free systems of education where everyone who tries hard gets equal praise, but in the real world competition rules. Those who get top grades have a good chance of getting key jobs. Those who simply 'try hard' but don't win have a tougher time. The thought therefore that you are slogging your guts out, while down the corridor someone is taking time out to party and then racing past with pill-powered learning is galling. It could even convince you that the only way to stay in the game is to take the tablets yourself.

Rumours abound about college kids illicitly taking Ritalin

As a society we value education highly, recognising that our brains need to be trained to think – or at least the more we train them, the more capable they are of thinking. Within a society that becomes important, as research shows that a country's gross domestic product (GDP), the sum of all wealth creation, is exponentially related to its people's mean IQ. This means that a small increase in average IQ is linked to a large increase in GDP. As a rough guide, a 10-point rise in IQ would be associated with a doubling in GDP. If you are planning education policy, that is a feature worth bearing in mind.

As we all know, caffeine has a strong track record of helping boost concentration over short periods of time, especially if contained in a drink packed with easily available energy. The caff-carb energy drinks pick you up and let you squeeze a few more productive hours out of the tube of life. But why stop with compounds that have been known for centuries? Why not try some of the newer chemicals on the block?

Joshua Foer's Adderall Adventure

One interesting candidate is Adderall, a close chemical cousin of Ritalin. Sometimes called the cognitive steroid, it has a track record of boosting concentration, alertness and attention. Writing in the US magazine *Slate*, Joshua Foer recorded his experience of giving it a go.

'As an experiment, I decided to take Adderall for a week. The results were miraculous. After whipping my brother in two out of three games of Ping-Pong – a triumph that has occurred exactly once before in the history of our rivalry – I proceeded to best my previous high score by almost 10 percent in the online anagrams game that has been my recent procrastination tool of choice. Then I sat down and read 175 pages of Stephen Jay Gould's impenetrably dense book *The Structure of Evolutionary Theory*. It was like I'd been bitten by a radioactive spider.'

Joshua said that the first hour or so of being on Adderall was mildly euphoric, but he found that that feeling wore off quickly, giving way to a calming sensation; 'like a nicotine buzz, that lasts for several hours'. When he tried writing on the drug, he said that it felt as if he had 'a choir of angels sitting on my shoulders'. He could pump out sentences without any sense of fatigue or distraction. Even the e-mail inbox failed to disturb his focus. While he could normally concentrate in 20-minute bursts, on Adderall he could stretch his mind for hours. The benefit came, not so much in his ability to feel smart, but to stick to the task.'

At the same time, he felt less like himself. Although Adderall let him write faster, he felt as if he was wearing intellectual blinkers. The focus was causing him to be cut off.

He is not alone in finding this. Other users find that what they gain in concentration they lose in creativity. It also comes with a pay-back time. Three hours after taking each dose, Joshua would start to feel groggy, but lying down for a nap had limited benefit because his brain was still fizzing. No chance of 40 winks. But did the benefits outweigh the harms? 'For me, the comedown was mild, a small price to pay for an immensely productive day', he reported.

> ❚❚ For **me,** the **comedown was mild**, a small price to pay for an **immensely** productive day ❚❚

Eat and Alter

The range of products that influence mood and mental activity is fairly large. Dopamine makes you awake and happy, even manic. Dopamine is built from the amino acid phenylalanine with the help of vitamins B6 and C. Eat plenty of these and you can boost your dopamine.

Compounds like lecithin, DMAE, piracetam, aniracetam and vitamins such as B5, B6 and C increase blood flow, and aid improved delivery of oxygen levels and other compounds. Alcohol, caffeine and other drugs affect our ability to relax or stay awake. Even activities like sport or extreme exercise increase levels of different neurotransmitters and improve mood, to the point that some people argue that sport is a mood enhancement.

The list goes on, and there are many websites and so-called health food shops where you can survey catalogues of supplements, each claiming some particular benefit. Using them can increase an individual's performance. Drugs that were originally intended to help people with conditions like depression or attention deficits can be used to boost performance in people who don't have clinical symptoms.

But could a drug be developed that would directly improve learning and memory retention – a 'smart drug'? Such a drug might have a wide market, from students studying for exams to the general population of over-50-year-olds suffering from mild cognitive impairment.

Research on the molecular mechanisms involved in memory formation has recently converged with attempts at treating cognitive decline in people with Alzheimer's disease. The result has been some new candidates. Drugs such as aricept and rivastigmine are designed to boost the neurotransmitter acetylcholine; memantine affects glutamate neurotransmission. Both can help people think more clearly. However, both have a number of unwanted side effects that make them less than instantly attractive.

The fact that the drugs give some benefit is enough to encourage further research. Some of it has drawn on our increased understanding of the way that the brain uses particular proteins when it stores memories. One new possibility comes with the acronym CREB. The hope is that the more CREB you have, the more enhanced your memory will be. Another company has focused on ampakines – a class of molecules that increase the power of certain receptors in the brain. Again, the hope is that enhanced receptors will give enhanced memory. Both seem to work in mice, but so far they have been less than successful when tried in humans.

One potential fly in the ointment for people wanting to create brains that are so capable they outstrip any that have gone before comes from research that gives drugs to people of different capability. The people who get most benefit are those who had low abilities in the first place. Those who could always think well received much less of an enhancement. On its own, this would suggest that there is an upper level of achievement – a glass ceiling – through which the human mind will not be able to pass. It raises the possibility that you could bring many more people to the point where they could think with the best, but that it is going to be much harder to stretch the biological maximum, to produce a brain that is distinctly enhanced.

You could bring **many more people** to the point where they could think with the **best**

Chemically Enhanced Armies

I'm sitting at my office desk hitting keys. I often work from home, where I have an office just off the kitchen. I've three young boys, but all is quiet. Peace at last. That's because it is gone midnight, and I'm still typing, thrashing keys against a deadline. My current deadline is a book – this one; others' are management-required reports or drawings that need finalising and signing off. What if I could pop a pill and have a consequence-free night without sleep? I reach for my coffee and discover it is half-drunk and cold. Time to pop it in the microwave and give myself a boost, but I know it will only help a little.

What if I could take alertness-enhancing pills at regular intervals and go for a week without sleep, remaining alive to what is going on at all times? Just think of the increase in productivity. The possibility hasn't been lost on the military. Going to sleep during combat is not a good way of staying alive, but combat doesn't fit into normal wake-sleep patterns. So there is a big push to find ways of extending endurance, or chemically enhancing our military personnel. One story, however, shows that with currently available drugs, this may not be as simple as it seems.

Friend or Foe?

Thursday, 17 April 2002, started much like any other operational day in Kandahar, Afghanistan, for the soldiers of the 3rd Battalion, Princes Patricia's Canadian Light Infantry Battle Group. They were engaged in a live-fire training exercise nine miles south of Kandahar airfield. They were practising an anti-tank mission, aiming weapons ranging from light firearms to shoulder-fired anti-tank weapons at inert targets in a recognised training area. It was all part of Operation Apollo, Canada's commitment to the US announced war on terror.

Canadian troops on patrol in Afghanistan.

The normal big-boys' wargame was catastrophically interrupted just after mid-night, however, when a pair of US military Falcon F16s sped overhead. The pilots, Major Harry Schmidt and Major William Umbach, reported seeing gunfire – action that Schmidt took to be enemy ground fire. He asked for permission from his commanders to mark the target and return for a second look. Permission was granted, and he was reminded not to attack unless he felt threatened. Flying back over the scene he once more saw fire, and again took this as a sign of being under attack.

Flying **back** over the **scene** he once more saw fire, and again took this as a **sign** of being under attack

Military might emphasises the power that one man has when he pushes a button. It makes 'successes' notable, but if something goes wrong, failure can have huge consequences as well.

'I've got some men on a road and it looks like a piece of artillery firing at us', Umbach said, according to a US military report. 'I am rolling in, in self-defense.' He released a 500-pound, laser-guided missile.

On the ground 29-year-old Sgt Marc D Leger, 25-year-old Cpl Ainsworth Dyer, 22-year-old Pvt Richard Green and 27-year-old Pvt Nathan Smith were at the centre of the blast. Another eight were close enough to be injured.

'The deaths, Canada's first combat fatalities since the Korean War, sparked anger among many Canadians, some of whom questioned their country's role in the American-led war on terrorism', said a CNN news report, and months later Schmidt and Umbach found themselves facing charges of involuntary manslaughter, aggravated assault and dereliction of duty. If convicted of all charges, they would face a maximum of 64 years in military prison.

The charge against Schmidt was that he didn't make sure he was dropping a bomb on the enemy, and he disobeyed air controllers' instructions to 'standby' while information was verified. The formal counts allege that he 'failed to comply with the applicable rules of engagement' and 'willfully failed to exercise appropriate flight discipline over his aircraft'.

Col. John Odom, who led the prosecution against Schmidt and Umbach, told a reporter from the NBC that the proper actions for pilots in that situation would have been to fly away and await orders.

Odom suggested that the evidence from the cockpit tape screamed that the actions were more like someone on the attack, not the defence. Schmidt denied this, saying that he was being set up as a scapegoat because he had not been informed that there were Canadians in the area.

In the end Umbach agreed to accept a reprimand and retire from the Air Force and the Air Force reduced the charge against Schmidt. On 6 July 2004, Schmidt was found guilty of 'dereliction of duty' in what the US military calls a non-judicial hearing before a senior officer. He was reprimanded and forfeited more than $5,000 US in pay. In the reprimand, Lt.-Gen. Bruce Carlson, who handed down the verdict, wrote that Schmidt 'acted shamefully . . . exhibiting arrogance and a lack of flight discipline'. The air force agreed to allow Schmidt to remain in the Illinois Air National Guard, but not as a pilot.

A High 'Speed' Attack

At best this is a tragic accident that is part and parcel of engagement in armed conflict – at worst grave criminal assault. One complicating feature, however, is that the whole incident could have been triggered by the fact that the pilots were taking dextroamphetamine. If you bought the drug on the street, you would call it 'speed'.

The basic chemical, amphetamine, was first synthesised in 1887 in Germany. Initially it was first considered as a drug to treat or possibly cure depression and other ailments. Since then, various alternative forms of the drug have been created, including the active ingredient of Benzedrine, an over-the-counter inhaler designed to treat nasal congestion. By the 1930s some elite athletes had discovered that it could boost concentration and endurance, and in the Second World War it was widely used by American, British, German and Japanese soldiers to keep awake and alert.

In the first Gulf War campaign, some of the pilots became addicted to the drug, and the pills were banned in 1993. Armed forces don't tend to shout about it, but use has crept in again, particularly by the few massively trained pilots who are expected to put in long hours in high-stress situations. Intriguingly, the pills would be illegal if sold on the street.

The use of the drugs is outlined in a 58-page document entitled 'Performance Maintenance During Continuous Flight Operations', written by the naval medical research laboratory in Pensacola, Florida. In its foreword, signed by R.A. Nelson, Surgeon General of the Navy, it says: 'Combat naps, proper nutrition and caffeine are currently approved and accepted ways flight surgeons can recommend to prevent and manage fatigue. However, in sustained and continuous operations these methods may be insufficient to prevent fatigue and maintain combat-ready performance. Properly administered use of stimulant and sedative medications, i.e. Dexedrine, Ambien, and Restoril, is an additional measure flight surgeons can recommend to manage fatigue and maintain pilot performance in continuous, sustained naval flight operations.'

A statement by the US Air Force Surgeon General's Office confirmed the use of amphetamines by pilots. It said: 'During contingency and combat operations, aviators are often required to perform their duties for extended periods without rest. While we have many planning and training techniques to extend our operations, prescribed drugs are sometimes made available to counter the effects of fatigue during these operations.'

The question is whether this use of amphetamines is simply a matter of keeping pilots awake. Or does it have less desirable effects? Commenting on this, John Pike, director of Globalsecurity.org, a defence think-tank, told the newspaper *The Independent*, 'When you look at the original story of the [Canadian] friendly-fire incident it seems that the pilot was being inexplicably aggressive. It goes beyond fatigue or lack of experience or [being a] cowboy or trigger happy or any of the standard prosaic explanations. The simplest explanation is that the guy had eaten too much speed and was paranoid.'

In addition, two unpublished reports into the friendly fire incident reportedly concluded that Mr Schmidt made his error because he failed to properly assess the supposed risk before striking.

The issue is that a recognised side effect of taking too much speed is a distinct feeling that everyone is out to get you. Add this to the understandable feeling that any pilot must have of being distinctly unliked and you could soon start to over-react.

Sam Patterson's Trip of Invention

I like to take my briefcase with sketchpad, cell phone (off), Pentel 0.9 mm mechanical pencil, flannel sheet, beach blanket, sun screen, and my "Zeppelin Smokeless Pipe" filled with the strongest marijuana I can find to a deserted beach.

I just make shapes. It doesn't matter. The mechanical devices I invent always have a definable input and output. There is also some design envelope that can be defined. I just sketch schematic notions of mechanical elements that solve some aspect of the problem. I don't try to solve it all at once. I just wander and watch and react as the sketches progress. Then sometimes I get a kinesthetic notion that something is coming. Sort of like a speeding train that you can feel and hear but can't see yet because it's around the bend. This is where I hang on and stay focused. In a brief flash a complete solution goes off. So fast that I don't quite register it all consciously, but I feel like it's somewhere in the subconscious buffer - however faint and fragile. That's when I start sketching like mad. If I'm lucky I can draw it out in the sketch. I don't fully comprehend it until I'm finished.

maps, volume X number 3, creativity 2000
http://www.maps.org/news-letters/v10n3/10320mem.html

Single Convention on Narcotic Drugs
http://www.unodc.org/pdf/convention_1961_en.pdf

Convention on Psychotropic Substances
http://www.unodc.org/pdf/convention_1971_en.pdf

United Nations Convention against the Illicit Traffic in Narcotic Drugs and Psychotropic Substances, 1988
http://www.incb.org/incb/convention_1988.html

Founded in 1986, MAPS is a membership-based non-profit research and educational organisation. MAPS assists scientists to design, fund, obtain approval for, and report on studies into the risks and benefits of ecstasy, psychedelic drugs and marijuana. MAPS' mission is to sponsor scientific research designed to develop psychedelics and marijuana into prescription medicines approved by the US Food and Drug Administration (FDA) and European Medicines Agency (EMEA), and to educate the public honestly about the risks and benefits of these drugs.

5 In Touch with the Brain

In 1984 William Gibson published *Neuromancer*. The timing was important. It was the Orwellian year when the world was looking out for possibilities of mind control, for opportunities of seeing Big Brother at large. The novel follows a computer hacker through a world where mercenaries are augmented with brain implants to enhance strength, vision, memory and a host of other features. It became the fore-runner of many books and films including the *Matrix* film trilogy, where people's bodies are used as energy supplies in a system where their brains are plugged into a huge artificial environment. It is a world where fantastic things happen – where people with super strength fight and fly with equal ease.

It is also a world that seems to be coming a little closer. Inserting long electrical needles into human brains is no longer the stuff of fiction, but in some corners of medicine, it is part of daily routine. There are now thousands of patients walking around with spikes in their brains wired to electrical stimulators that are usually slipped just beneath the skin of the chests. There are also a few people who are beginning to control computers just by thinking, using sets of electrodes stuck to their skulls.

Pausing Parkinson's

To find out what you could actually achieve with spikes implanted in the brain, I phoned Professor Tipu Aziz, a pioneering neurosurgeon who works at Oxford Functional Neurosurgery. He has developed a reputation for tackling movement disorders such as those seen in people with Parkinson's disease and I soon found that it went much further.

Parkinson's disease is nasty. The sufferers gradually develop uncontrollable shaking movements that take over life. Various therapies have been tried and had some success. Carefully burning away specific parts of the brain can turn off the

Based in Oxford, UK, Professor Aziz is one of the world's most experienced surgeons when it comes to inserting stimulating electrodes in the brain. The results can be impressive, but it is still brain surgery and carries significant risks.

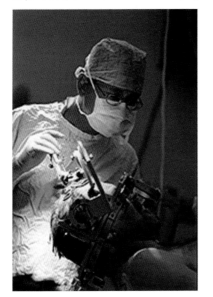

symptoms, but the procedure comes with a risk of altering a person's mood and damaging other brain functions. Another option is to give a drug called L-dopa, but this again has problems. Some people develop symptoms of psychosis, others an uncontrolled writhing that is in many ways worse than the original shaking. A series of discoveries linked to a group of drug addicts who had accidentally given themselves Parkinson's-like symptoms, led to the discovery that the tremors were removed if you knocked out an area of the brain called the sub-thalamic nucleus, or STN.

Carefully **burning away** specific **parts of the brain** can turn off the symptoms

Removing the STN is extremely dangerous in patients because the area is packed with blood vessels and the operation can cause massive bleeds. This would leave patients with even less control of their movements. Thankfully, French neuro-surgeon Professor Alim-Louis Benabid discovered in 1987 that if you implanted a wire in the patient's STN and stimulated the area, the tremor stopped instantly. It was a huge breakthrough.

The more we find out about the structure of the brain, the more possibilities there are for treating brain disorders. From there rises the possibility of adding new capability.

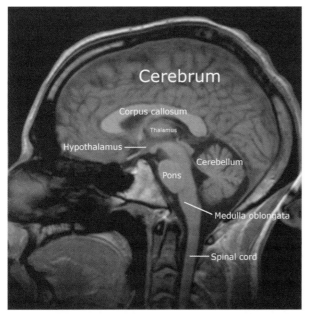

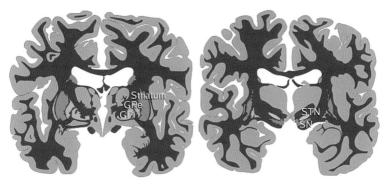

Deep brain stimulation has been used in the treatment of movement disorders, such as Parkinson's disease.

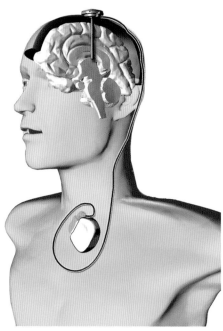

Want to see the power of an implant? Then view these videos:
www.dana.org/events/detail.aspx?id=7326
http://www.guardian.co.uk/science/video/page/0,,2139463,00.html

In it some researchers and surgeons, including Professor Aziz, explain the history and science of deep brain stimulation, and Parkinson's patient Mike Robins provides a stunning demonstration of the power of the implant at work.

In the years since 2002 there has been a rapid increase in the number of people receiving the implants, and now over 14,000 people worldwide have symptoms reduced or removed by electrical currents passing through long spike-like needles deep in their brains. Quite how it works remains uncertain. The general view used to be that the current was likely to block or confound nerve activity in the area, rather than stimulate it. New research, however, is suggesting it may stimulate some areas as well.

Wonderful though it is, deep brain stimulation hasn't cured the patient; he or she still has the underlying problem. The technique therefore falls more into the category of a really effective walking aid, or crutch rather than a treatment. It certainly isn't an enhancement, even though it does let life resume for the needy people who have it.

A Controlled Turn On

Parkinson's, however, is not the only use for this technique. It has now been used to control symptoms of depression, obsessive-compulsive disorder, Tourette's syndrome, chronic pain and cluster headache. Interestingly, stimulation can occasionally cause side effects such as depression or further movement problems, but the side effects can also include mirthful laughter and penile erection.

I asked Aziz whether he would describe any of the currently achievable effects of stimulation as an 'enhancement'. He started by thinking about patients who have deep brain stimulators installed to combat pain.

'What patients report to me as a result of having this surgery is that it's not so much that the pain is removed, but that what is removed is their emotional reaction to pain', he replied.

This description seemed similar to the response that people in pain feel when they are given a painkiller based on some opium-like chemical. They say that the pain is still there, it just seems to be a long way away and not part of themselves. It is therefore much more tolerable. But then the technique can go one intriguing step further. Professor Aziz believes that you can extend the idea and stimulate the same target in someone who wasn't in pain. The result could be that they would enhance the intensity of their emotional life, or possibly make them less likely to feel sad.

Now some people might pay good money to be able to instantly switch off depression or trigger mirthful laughter. And if the plethora of adverts that pour into my junk email bin each day is anything to go by, there must be a large market for blue pills that enable men to control erections. Just think of the possibilities of an implant that gives you control over your moods, enabling you to turn on and off at will.

Just **think** of the **possibilities** of an implant that gives you control over your moods, enabling you **to turn on and off at will**

Just think. It's been a hard week and things have not gone well. The evening looks promising, but you are not in the mood. You are depressed, anxious and feel that your libido is low. Simple. Reach out and grab the mobile phone-sized controller and set it to 'partytime'. Hold it over the small box of tricks implanted under the skin on your left chest and press 'send'. A train of electrical pulses streams into your brain, and problem solved. From worn out office bore to confident bubbly party beast in a few seconds.

But am I getting carried away and beginning to slip into hype myself? Aziz doesn't think so. Systems not vastly different from this have already been installed in people with mood disorders and in some of these people they work well. Right now they don't work on everyone who has them installed, but Aziz believes that that is probably because they don't yet know the right place to target in the brain. The more patients they fit with the devices, however, the better their understanding of brain function. 'So we can, theoretically, think about enhancing the mind, or one's perception of life, or whatever you call it. I mean, what do you want to do? Do you want to feel less pain by life's tribulations, in which case, you could choose one area of the brain to target. Do you want to be free of depression? You choose another target. These are theoretical possibilities', he comments.

Aziz is excited about using a new imaging technique that will allow him to scan a person's brain at the same time as turning on the stimulator. Until recently the most effective form of brain scanning was MRI. This involves placing your head inside a massively powerful magnet that makes all the atoms in your head line up on their magnetic axes. A pulse of radiowaves then knocks them off balance, causing them to spin, and sensors can detect that spinning. Different types of molecule spin in different ways, and using this information a computer can build up a picture of what is going on.

The problem with this is that if you have any metal in your head, then the magnet is so powerful it will pull it out. Best not go there with a brain implant then.

A new alternative, however, is a MEG scan. This monitors the natural magnetic activity inside the brain, rather than using an external magnet to create an artificial one. The results may not be as clear as an MRI scan can achieve at the moment, but for researchers like Aziz it is a great step forward. They can now stimulate one part of the brain, and see what happens right at the tip of the electrode, and also whether other parts are affected.

Magnetoencephalography (MEG) scans show which bits of a person's brain are active at any one time.

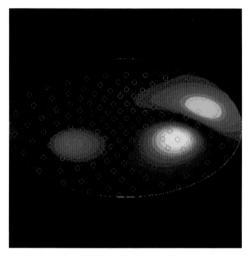

For Aziz, the question arises of whether a person with a normal brain, living a normal life, would want to undergo the moderate risk of brain surgery to gain this sort of enhancement. I think that by the time you have met some of the risk-takers in Section III, you will give an undoubted yes to that. Controlled brain surgery could almost be considered a minor risk relative to some of the risks people already take to push their bodies' boundaries.

A Sporting Chance

And how about other areas of life? If you are competing in rifle shooting or snooker, it is very important to keep still and not to shake. After all, everyone's got a tremor of some sort, and it would theoretically be possible to eliminate it with implanted electrodes. Just think of a surgeon who's wanting to perform a feat of fine surgery – a click of a switch could be a great way of steadying the hand a few hours into a complex procedure. But then, Aziz points out, we need to balance risk and reward. No surgery is risk free and any form of brain surgery is always undertaken as a last resort. You need to be in severe need before the trade-off makes sense. Certainly nothing in Aziz's tone suggested that he was about to rush out and get an implant himself in the near future.

Listening In

I went to Aziz without any particular expectation to find enhancement. From my starting point it seemed that his technique was a form of highly controllable knock-out surgery. He could specifically disable bits of the brain, and thereby remove unwanted movement or pain. That fits into a model of therapy, but less well into enhancement. But finding that there is real potential for controlling mood, not only in those who are clinically depressed but also in people who would like a recreational kick, takes the technique into a new direction. You can't buy it off the shelf yet, but a happiness switch is not something I or anyone I know was born with.

There is **real** potential for controlling mood, not **only** in those who **are clinically depressed** but also in **people** who would like a recreational kick

The next person on my list I approached with more expectation of finding enhancement capability. While Aziz is turning off unwanted functions, there is an increasingly large group of scientists around the world interested in monitoring brain activity, and using the signals they pick up as a novel way of letting the brain tell us what it is thinking. In 2006 the front cover of *Nature* magazine featured a man who had no ability to move his arms or legs, but who could open and close a robotic hand just by thinking.

Go back a couple of years and in 2003 *New Scientist* magazine ran a story about Hans-Peter Salzmann, who had learned to type on his computer just by controlling brain waves. Hans-Peter has Lou Gehrig's disease, or amyotrophic lateral sclerosis (ALS). It affects between 1.5 and 2 in 100,000 people and gradually destroys all voluntary movement. His symptoms have developed to the point that his mind is effectively locked inside a paralysed body. He can't even breathe for himself – a respirator does that.

But he can think – and can use this to communicate via what the scientists call a 'Thought Translation Device' (TTD). Two electrodes are stuck to his scalp to measure the electrical signal produced in the cortex immediately before a thought or an action – in Hans-Peter's case, he is limited to thoughts. These signals are fed into a computer and let him move a cursor.

The software lets him choose letters one at a time by making a series of yes/no decisions. For 'yes', he leaves the cursor alone, for 'no' he causes it to move. To move the cursor he tries to build up tension in his mind by concentrating on images like a bow being drawn or traffic lights changing colour. Then he releases the tension and lets the tension explode by letting the arrow fly or the light change. Typing is not fast – but it is possible.

These sorts of stories have been combined with futurist exuberance to talk of the immanent arrival of fighter pilot helmets that can take the aircrew's thoughts and use them to fly the machine, unleash weapons or deploy ejector seats. They creep into many a sci-fi movie, making them seem palpably present.

So I was a little surprised at the opening comment from Stephen Roberts, who works in Oxford's Pattern Analysis and Machine Learning Group, and, amongst other things, develops ways of listening to brain waves. I'd explained that I was writing the book because I was fascinated by the science and the discovery of the way our bodies worked, but unsure about whether we were actually about to produce an enhanced human and so I was calling on those at the cutting edge and asking them to convince me.

'The first thing is I really, really share your scepticism about enhancement', he said, to open the conversation. He explained that the main thrust of their work is to produce communication devices for people who have problems with standard ways of communicating. They might find ways of helping someone use a blow pipe instead of a computer mouse, or of getting cameras to track eye movements.

The interesting thing is that the electrical activity in the brain is the same if you move your hand, or if you just think about moving your hand, but can't because it is paralysed.

This is where other parts of Roberts' work cut in. He and his team have developed ways of finding information within noisy signals. They have built a system that uses just two electrodes. Other people use up to a hundred electrodes, but Roberts is keen to make the technique as practical as possible. It takes two hours to stick on loads of electrodes; two only takes a couple of minutes.

According to Roberts, picking up the brain activity is like trying to listen to one person in a cocktail party of a hundred. He heads up a team of information theorists rather than neurophysiologists, which has an implication in the way he tackles the work. While neurophysiologists are used to the idea of neurones having many connections, and therefore try to mimic this by using many tens of electrodes, Roberts uses just two. The technique doesn't limit the amount of information he can collect, but it does demand a lot more computational analysis. His argument is that he prefers to spend time teasing the signals out, while other people spend time sticking electrodes on.

Picking up the **brain activity** is like **trying** to listen to **one person** in a cocktail party of a **hundred**

Using brain activity to trigger an action currently relies on one interesting feature of the way brains work. Just before an able-bodied person performs any movement, such as wiggling his or her fingers or walking, the primary motor cortex in his or her brain enters a movement-planning phase. If the person closes his or her eyes and imagines the action, the brain still performs the movement planning.

This has been known about for years. What is new is the ability to pick electrical activity up with stick-on scalp electrodes. Roberts' group is developing computers that pick out the patterns of activities in individual people as they think about movements. In his system, the patient has an idea of how well he or she is doing because a moving cursor can be seen. This lets the computer learn about the person and the person learn about the computer.

This human–computer interaction is often referred to as a kind of symbiotic learning. Roberts is excited and cautious about the concept in equal measure. The work is important and he is aware that it sounds really exciting and very cool; the term 'symbiotic learning' certainly goes down well when writing grant proposals. But in practice he is only too aware how far there is to go. If you get to the nitty-gritty, the kind of communication rate we're considering is about one bit of information a second. Now a bit is simply a yes/no answer. You can do that speed robustly but you really can't push faster.

If you imagine playing a yes/no answer game, such as a form of 20 questions, you can go a long way in 20 seconds. But in order to fine control the joystick on a wheelchair, for example, you actually need a lot, lot more information than just a little yes/no at once a second. Consequently, Roberts believes that in terms of some of the more practical aspects, there are fundamental limitations to the use of this kind of technology.

Roberts too had seen the *Nature* article describing the man who could open and close a prosthetic hand by controlling patterns of brain activity. He was cautious, though, in extrapolating from that to suggest that we were on the verge of having mind-controlled machines, much less seamless connections between mind and mechanical matter. For him, the technology in that work created little more than an on/off switch – you could, after all, change open and close for on and off. 'If they had replaced the artificial hand with a light that went on and off, it wouldn't have looked nearly so impressive. Now, you can put a lot of intelligence in a prosthetic limb, and you can initiate a grasping motion and so on, but my goodness, we have so far to go before that grasp becomes useful. There's so much hype in this whole field and we've a long, long, long way to go', he comments.

One issue is accuracy. At the moment the best you can achieve in terms of getting a computer to follow someone's yes/no thought sequence is around 70%. Three out of 10 occasions the computer will get it wrong. Now that may seem good, because it is better than 50%, but you wouldn't want to let someone loose in a car with that accuracy.

The only way to get around the accuracy problem is to do things very slowly. Roberts says that the take-home message is that you can do billions of times more with a little finger that can wiggle consistently than you can by putting skullcaps on. He does acknowledge that there is a small population who have absolutely no controllable motion, and they may benefit, but he insists that if you can move your eye or blow into a tube you will be far ahead of systems that rely on monitoring the brain. He is less than enthusiastic about suggestions that pilots could use these sorts of systems to fire ejector seats and deploy a parachute in high-G situations. As far as he is concerned, you could build a much more reliable system by getting someone to blow in a tube, scream or exhale – any of those would be better triggers to use and would be more reliable in a stress situation than relying on an electrode-filled skullcap.

Taking a cold hard look at this area, it is clear that there is some fascinating science going on and we are learning lots about the brain, but that there is a huge gulf to cross before this research generates a powerful enhancing tool. Getting around the problem of moving from a not-very-accurate yes/no type of response to an accurate proportional response, so that you can get a brain-controlled system to accurately move a cursor partway across a screen, is not going to be easy. That sort of fine control doesn't exist right now. You can get a little way with implanted electrodes, where you look at the responses of individual neurones, but then you can only keep the electrode next to an individual neurone for a few minutes, so again not a very useful tool.

This is a long way from developing a technology that will enhance human existence. We are learning much more about the brain, how it executes movements and how humans and computers can learn to learn together. But it would be foolish at the moment to assume that this is rapidly going to lead to a mind-controlled fighter plane or even a mind-controlled, dexterous artificial hand. We are just touching the tip of the iceberg at the moment.

'I don't want to sound too negative, because I mean, you can do a lot as I say, with a yes/no question every second. But we are nowhere near the information content of a few words in a sentence or a small image or a little bit of speech', concludes Roberts.

Enough said – I think.

6 Maxing Memory

The CD whirrs in its cradle and the unmistakable voice of Barbra Streisand cuts through the air: 'Memories . . .' Not only do the lyrics spring to mind, but so do the images of the school disco where I heard if first. They flit like shadows across the mind, like a badly damaged videotape of frozen images and short sequences of action. I can even conjure up the smell of floor polish, sweaty bodies and over-heating dusty disco lights.

I went back to my old school 18 months ago. It was the first time I'd been there for almost a quarter of a century. The hall was unchanged – same floor, same stage, same curtains . . . but it had shrunk. As a schoolboy it was the largest room I knew – now it seemed to be a rather small space, one that no longer matched my recollection.

It's well known that our memories play tricks with us. In a scientific paper I read recently, researchers reported showing photographs of the Tiananmen Square massacre to one group of people. A second group was shown another version of the photograph that had been doctored to give the impression that many more people were present. They didn't let on that some of the photos were fake. The researchers then asked questions about the incident itself. The version of the photo had a significant influence over how people answered – it had altered what people 'remembered'. Interestingly, it also altered their responses to questions such as whether they would join in a similar protest if it occurred in the future.

Stretching What We Have

Enhancement enthusiasts frequently talk of memory in the same sorts of terms as do computer scientists. It's not entirely surprising, as many of them come from computing and engineering backgrounds. As a consequence, the answer to fickle memories is to improve our capacity to store information, and our ability to recall it correctly.

Ben Pridmore **from Derby**, England, holds the **world record** for **memorising the sequence of cards** from a single **pack – 26.28 seconds!**

Most of us, however, hardly even start to use the memorising capability that we are born with. We are shocked at the abilities of people who memorise lists of random words, or the sequence of multiple decks of cards. The pinnacle of achievement in these memory games is the World Memory Championships, which in 2007 were held in Bahrain. Over the three-day memathon competitors stepped up to perform many different tasks that tested different aspects of memory.

In one feat they had to see how long a string of binary digits – zeros and ones – they could memorise in 30 minutes, with 60 minutes given for recall. The winner of that event was Ben Pridmore from Derby, England, who achieved a world best by recounting a 4,140-long sequence of ones and zeros. He also holds the world record for memorising the sequence of cards in a single pack – 26.28 seconds. Not bad – just you try to say the card names in half a minute!

In another event, competitors are given one hour to memorise the sequence of playing cards in as many shuffled packs as possible. They then have two hours to write down the sequences of each pack. The scoring system is tough. If you get all the cards in a pack right, you score 52. One mistake and you only get 26. Two mistakes and you score zero for that pack. Gunther Karsten, a memory grand-master from Erfurt, Germany, won this category with a personal best score of 1,044. He went on to win the World Championship as well.

Can Anyone Do It?

So are these memory giants freaks or just people who have practised hard and developed skills that most of us seldom even start to put to work? Joshua Foer, who we met earlier in the book trying Adderall, also became fascinated by what people could remember after attending the US memory championship as a journalist. He first decided to write a book about their techniques and secondly signed up for the following year's championships.

Memorizing sequences of cards rests on creating a story out of apparent chaos. Could an enhanced brain do better?

One basic technique is to build a virtual living space in your imagination and leave the objects in specific places. You then practise walking through that space in a set order and can pick up the objects as you go. A classic journey to take is a golf course. Many players can describe rounds of golf that they played years ago, recounting each stroke, the club they used and the place the ball landed. If you are trying to memorise a sequence of cards, pick one of your favourite rounds, and place the first card next to the first tee. The second card can then sit next to the club you pulled from the bag, and the next card on the fairway next to where the ball landed. Even if you only took three strokes for each hole, you would position every card in the pack well before you reached the eighteenth tee. The next pack can then piggyback on another memorable match.

Joshua took a slightly different tack and placed objects in his virtual world in distinct places around his home. He starts by taking a walk through his home. He

The hope of enhancement is a multitasking mega-brain. The hype is that we are about to achieve it.

places the first card next to the front door and gives it a character, say former US President Bill Clinton. He visualises Clinton at the door and then moves on. He walks into his house, places the second card behind the door and translates that one into a horse. Bizarre? Well that is part of the plan. For him, and many others who have tried it, the more bizarre the images, the more they stick in the mind.

It strikes me that this creates one of two challenges for enhancement buffs – the first option is to say that an enhanced memory is one that can do better than these apparently mind-blowing feats, and the other is to say, 'let's find a way of enhancing people so that they can achieve this human maximum with the minimum of effort'. This alone would enhance human average capability and have interesting impacts on society. One of the problems in security is remembering PIN numbers. Imagine a world in which people had been trained to remember 100-digit sequences – just think how much more complex your credit card security could become.

Looking for Boosters

So if we want to make retaining information easier, we need to look at what happens in the brain and see if we can enhance it.

One interesting thing is that strong emotions tend to make strong memories, and now there is some good science to show why that might be the case. It seems to be linked to the fight, flight, fright response – our adrenaline-driven hormonal reflex that lets us sprint when in danger. When going through an emotionally charged experience, adrenaline washes through each person's amygdala, an almond-shaped part of the brain that processes emotions like fear. Adrenaline tells this area to wake up and take note.

Strong emotions tend to make strong memories

We now have a handy arsenal of chemicals that prevent adrenaline from working. Called beta-blockers, they prevent the hormone being recognised by the cells that normally respond to it. One of them, propranolol, has raised particular interest. In research carried out at the University of California in Irvine in the late 1990s, a group of people were told an emotionally neutral, comparatively boring story illustrated by 12 slides. A second group was shown the same 12 slides, but this time the story they were told was much more emotional, involving a severely injured boy.

Later they were asked to say what had been in the pictures. Those who had had the emotionally charged story could recall the pictures in much greater detail than did the first group. Enhancing memory could therefore be a matter of working under controlled stress.

But then they turned this idea on its head. The researchers got a third group of people and told them the emotionally upsetting story while showing the same set of slides, but this time they gave them a standard dose of propranolol or another beta blocker. Once again their memories of the pictures were tested three weeks later. Even though they had been told the high-emotion story, their recall was just like that of subjects who had received the boring story.

The potential for this has not been missed when it comes to helping people cope with traumatic incidents. The researchers went in to the emergency room at Massachusetts General Hospital and found 31 people who had just been involved in a large car crash. Around half were given propranolol for 10 days, and the other half were given sugar pills that looked the same. That way no one knew whether they were taking the drug or a placebo. One month later none of the propranolol-treated people had any stressful reactions when the incident was replayed, while six out of 14 of those on the placebo were troubled.

Enabling the brain to chill out meant that it had failed to take note of the incident, and subsequently the person could be much more detached and less affected.

Not everyone is convinced that this enhanced amnesia is necessarily a good thing. A report from the President's Council on Bioethics worried that the 'use of memory-blunters at the time of traumatic events could interfere with the normal psychic work and adaptive value of emotionally charged memory. A primary function of the brain's special way of encoding memories for emotional experiences would seem to be to make us remember important events longer and more vividly than trivial events.'

Strong Connections

For decades it has been accepted that memories form as the strength of connections in the network of neurones in the brain increases. This is done in two ways; firstly by increasing the numbers of branches that individual neurones send to other cells, and secondly by increasing the number of chemical receptors that a neurone builds at any one junction. One of the key players is the NMDA (n-methyl-D-aspartate) receptor.

In 1999 a team of researchers in the USA published an intriguing paper in *Nature*. They set out to see if adding genes to mice to make them more capable of building NMDA would lead them to have better memories. The researchers reported: 'These mice exhibit superior ability in learning and memory in various behavioural tasks . . . Our results suggest that genetic enhancement of mental and cognitive attributes such as intelligence and memory in mammals is feasible.'

Life's Painful Lessons

So at first sight there is hope for someone who is looking for an enhanced memory fix. But before you rush out to get enhanced, the second half of this research paper bears careful examination. The genetically enhanced mice were tested in a variety of different situations. On occasions they were allowed to explore spaces and objects, and on other occasions they were put in chambers and subjected to a combination of loud noises and small electrical shocks to the feet. The researchers were able to see which mice remembered the nasty situation most clearly by seeing which exhibited fear most strongly when put back in a similar chamber on subsequent days.

The memory-enhanced mice indeed remembered better. Mice tend to freeze when worried, and the modified animals were much more prone to have a fear response. The best explanation is that the unmodified mice had forgotten that some situations caused distress and wandered on in blissful amnesia. The enhanced mice remembered it only too well, and were haunted by the prospect of meeting the pain again. Far from enhancing them, it could be seen as inhibiting their actions.

Unmodified mice had forgotten that some **situations** caused distress and wandered on in blissful amnesia. The **enhanced** mice remembered it only too well, and were haunted by the **prospect** of meeting **the pain** again

This research shows that pain, learning and memory are linked, so perhaps the adage tripped out by schoolmasters down the years that there is no gain without pain has some basis in science. It also makes evolutionary sense, in that you need to learn to avoid nasty things, and then remember those lessons, if you plan to stick around for a long time. There is nothing like picking up a hot twig that has dropped out of a bonfire to teach a child that fire is hot – once done, twice shy – it is a lesson that sticks with most for a lifetime.

Selective Amnesia

The ability to manipulate memory is an idea played with in the 2004 film *Eternal Sunshine of the Spotless Mind*, in which a brain scientist has supposedly conjured up a mechanism for selectively erasing individual people from your memory. It's not a new idea – In Shakespeare's *Macbeth*, a doctor is asked to come up with an antidote that will clear Lady Macbeth's memory-troubled mind.

'Canst thou not minister to a mind diseas'd,
Pluck from the memory a rooted sorrow,
Raze out the written troubles of the brain,

And with some sweet oblivious antidote
Cleanse the stuff'd bosom of that perilous stuff
Which weighs upon the heart?'

But time has moved on, and while there was no hope for Lady Macbeth, there are possibilities on the horizon. Like much good fiction *Eternal Sunshine of the Spotless Mind* had a take-off point in recent science. Just as there are scientists working on memory enhancement, there are others seeing if they can selectively wipe some of it away.

Memories are complex, partly because different places in the brain are involved in different aspects of the task. Some areas work on short-term memory. For no more than a few seconds, they hold information and numbers that you need now, but can forget about in the future. Other areas are employed in holding the archives, the long-term memories that become more ingrained the longer they stay in place.

How happy is the blameless vestal's lot!
the world forgetting, by the world forgot.
Eternal sunshine of the spotless mind!
Each pray'r accepted, and each wish resign'd;
From *Eloisa to Abelard* by Alexander Pope

What is interesting, though, is that when you recall a long-held memory you lift it out of the archive and place it in the short-term zone. You can now play with it, interrogate it, and it in turn can influence your thoughts and actions. When you have finished, the biological equivalent of a librarian comes and packs it up again and files it carefully away in the archive. Without this re-filing, the memory would be lost. The question is, what would happen if you could kidnap the librarian and prevent her from acting? Would the memory be lost?

One research group believe they have done just that. They started by training rats to fear two different musical tones, by playing them at the same time as giving the rats an electric shock. If they played the music on its own, the rats got frightened – they could remember that the music normally came with something bad. The researchers then gave half the rats a drug known to cause limited amnesia. It's called U0126, and it is not available over the pharmacist's counter. They then replayed just one of the sounds, causing the rats to recall just one of the fearful memories.

A day later they retested the rats with both tones. The untreated animals still feared both sounds – they appeared to expect an imminent shock. Those treated with the drug, however, were no longer afraid of the tone that had been played while they had the drug in their system. They did however appear to panic when the second sound was played. The most obvious explanation was that recalling the memory while U0126 was floating around the body caused it to be selectively erased.

Blessed Are the Selectively Forgetful

The question with most technology, however, is learning when to use it. In *Eternal Sunshine of the Spotless Mind* most of the people whose memories were spring-cleaned found that the process was not as beneficial as they had hoped. This was clearly the case for the senior scientist and his secretary. The memory of their previous affair had been erased from her mind, but that meant she was no longer in a position to learn from her mistakes and so found herself drawn to repeat the act. The point being made was that sometimes even painful memories are important.

> Blessed are the forgetful: for they get the better even of their blunders
> (From *Beyond Good and Evil* (1886) by Friedrich Nietzsche)
>
> I'm not sure that most people with Alzheimer's feel all that blessed. It's a question of choice – if you can choose to forget, then you have a useful tool.

In other situations, there can be little benefit from being plagued by constant recall of chapters in our history. There would be many people who suffer from post-traumatic stress disorders who would find that that sort of selective memory erasure could distinctly enhance their lives. Does this mean that they would be enhanced humans, or that they would have been humans whose lives are enhanced by good therapy? The answer depends on your definition of enhancement.

The Stuff of Individuals

Debate about memory or forgetfulness enhancement goes again to the heart of what it is to be me. However close their similarity at early stages of development, identical twins certainly cease to be identical as soon as they start to experience life for themselves. Each has a unique set of memories and builds his or her responses to life's challenges in response to that stored history.

To change the memory bank will go a long way to changing the person. Leon Kass, who chairs the President's Council, complains that any drug that could erase memory would remove the need for regret, remorse, pain or guilt. He says that the requirement of good technology is to help people not to fall apart in a crisis, rather than transform them into something different. If the horrible things in the world no longer disturb you, if you effectively dull your senses, you 'would cease to be a human being', he says.

In that case, would you have transcended humanity and become more than human, become post-human, or fallen behind and become a mutated version of the species? The issue could be one of control. If you decide, having experienced the horrors, say, of war, and need to move on in life, then dulling the mental pain could be one way of re-entering civi-street. It would be your choice and your benefit.

What if your military commanders insisted you took a 'pill' so that you forget the awful horror of a previous mission in which friend and foe were blown apart before your eyes? The amnesia would free you to enter the next mission unencumbered by flashbacks or remorse. You would, I believe, become less human and more a cog in the military machine.

III   Better than Life

A man chases through dense woodland, keeping low, panting. It's dark and he trips over broken branches, tumbles into moss, struggles to his feet, looks over his shoulder and pushes on, trying to get away. The camera jolts around as we get an impression of the task he has to face, the impossibility of seeing, the difficulty of staying upright. Another man runs with a more measured pace and surer foot seemingly able to dodge obstacles at will – he is closing the distance on the first, and in contrast seems unperturbed by the exertion. The view changes and it is clear we are now seeing the world through his eyes. It looks different. Superimposed on a clear, stable and monochrome view of the forest are distance markers, and gun sight marks. The view clouds slightly, but a blink and its clarity comes back. He stops, and the muzzle of a gun comes into view. It is tremoring slightly as he lifts and aims, but then suddenly the tremor is gone, the view and the gun become fixed and our runner falls to the ground. We now see the pursuer's face. It doesn't move but we hear him confirm the kill to some distant voice. [camera fades to black]

It's the stuff of science fiction, or rather increasingly the stuff of fiction that stretches that little bit so that the boundaries are lost. The result is persuasive as you find yourself drawn into the normal experience and then start to see the abnormal in the same light. You lose a sense of what is immediately possible and what is a script-writer's imaginative genius. These are great films for practising 'what if' debates. What if you could see in the dark with added distance ranging sonar built in? What if you could stimulate areas deep in the brain that instantly prevents tremors? What if you could run and run without becoming breathless?

What would a society look like if some people had these capabilities and others didn't? What if no one asks these questions and we stumble on this situation unprepared?

It's not just the realm of movies that creates this imaginative crisis. I watched a BBC documentary in annoyance a few years ago. It was a flagship series unveiling the wonders of the human body, and at this point we were exploring the ear. It is amazing, said the narrator, how much we can now see. Cameras are so small they can go right inside the ear. And with this we swept into the outer ear and had a fascinating view of the eardrum. I winced, however, as the camera didn't stop. I expected it to bounce off the drum and hear some statement about the closed nature of this outside part of the ear. Instead it continued through the drum as if it were porous and then more remarkably travelled on up the inner ear, a journey that would defy any camera.

What had occurred was a wonderful piece of sleight of hand – we had emerged from the world of real video and seamlessly entered video graphics. This was no longer a camera view – it was a composite built by scanning and coloured by imagination. You could argue that getting the narrator to say that we now look at an artist's impression of the rest would lessen the impact. True. But as it stood it left viewers with a false impression of what technology can actually achieve.

In the real world, a few people have attached items of technology to their bodies in order to try and build better humans. Even more have been talking about these possibilities, but then tend to merge seamlessly into talking about potential ideas that no one has ever tried yet.

To an extent, there is nothing new about modifying bodies. Ancient tribes have done it for years, as did the Chinese when they bound women's feet ensuring that they were incapable of growing. Others have made holes in lips or tattooed exotic patterns over themselves. What I am interested in now is what technologies people have built in, as they attempt to give a human body new functions – as they have tried to make it better than life.

7 Conformity in Enhancement

It's not every day that you get the chance of meeting someone who has permanently added a new method of sensing the world to his or her body. So I was excited at the possibility of meeting Todd Huffman. He is a colourful character who enjoys dressing up to cause a stir, and if he feels the occasion allows, is just as likely to turn up wearing a skirt as trousers. It's nothing about sexual identity, but has a lot to do with attempting to push boundaries and challenge expected norms. It's about allowing absurdist experimental art and performance art to enter everyday life.

What intrigued me most, however, is that Todd has a magnet implanted in his left ring finger. Todd had started out as a trainee nurse working on neurology wards in St Louis, Missouri. He discovered then that he didn't want to be a doctor but was interested in medical science, so he signed up for a neuroscience Bachelor's degree. He transferred to Long Beach, California, for the last part of his studies. Developing an interest in transhumanism, Todd moved to work for Alcor in Scottsdale, Arizona. That's Alcor Life Extension Foundation, one of the two famous American cryonics companies that freeze people when their hearts have stopped but preferably before their brains are dead. In the hope that some future technology will allow them to be thawed, reanimated and cured. Since then his interests in extending human capability have grown year on year.

Substance, Sense and Structure

I met up with Todd at Paddington Station, London, and along with his girlfriend headed for a meal in a nearby Italian restaurant. To an extent, I was a little disappointed that he had turned up in jeans – I thought the dress could have been an interesting starting point for discussion. Then again, I think the skirt, costume, whatever, would have been a distraction.

Pushing boundaries for Todd includes not only adding a magnet to one finger but building a small threaded stud into his cheekbone.

© PETE MOORE 2007.

I was, of course, particularly intrigued by the idea of having a magnet implanted in a finger. I had a whole stream of questions – what, when, where, why? It turns out that the implant is quite small – about the size of a thick grain of rice. The magnetic part of it is made of neodymium, which was then plated in gold and encased in medical grade silicon. It had been implanted on 24 January 2004 into the pad of the fourth finger of his left hand, just to one side of the finger bone.

Neodymium is classified as a rare earth metal. It comes in at number 60 in the periodic table and when combined with iron creates the powerful permanent magnets used in microphones, speakers and guitar pickups.

Todd's implant was inserted by Steve Haworth, a US-based body artist, who has a lot of experience adding elements to people's bodies, ranging from silicon implants that give three-dimensional structure to tattoos, to stainless steel horns bolted to people's skulls. Todd knows Steve and his work well, as he used to live in a warehouse with a bunch of other people who were into body modification.

Web Link

At www.stevehaworth.com, you can get a fuller idea of what body modification can mean.

The idea had come during a conversation with one of Todd's transhumanist friends to test the idea of substrate independence. This theory says that we are used to understanding that, for example, our brains are the places where thinking happens. We are also getting greater knowledge of how the intricate network of neurones enables this process. But what if we could build some other system that can achieve the same thing, but uses a very different material – a different substrate? If the new substrate enabled the same thinking to occur you could say that thinking and possibly intelligence were 'substrate independent'. Similarly, if Peter Houghton's constantly spinning rotor pump (see Chapter 1) were ever able to perform as well as a natural one, then heart function could also be described as substrate independent. Whether it is really independent, however, is still an open question. Many philosophers of mind, people who think about thinking, have grave doubts.

Enhancement and Substrate Independence

Substrate independence is an important concept in much of the thinking around enhancement. If thoughts can occur in a microchip, as well as in a brain, then thoughts are independent of the substrate that supports or enables them. If that is the case, then there are many possibilities for re-thinking our thinking. You could set out to build a brain that was more robust, had more memory, lasted longer or was simply renewable. Also, if memory and thought were both substrate independent, then uploading would be very much easier. Another option is that you could add upgrades and extensions to people as they used up all the 'space' that they have to start with.

So how does this bring us to a magnet in the fourth finger of your left hand? The cells in our brain are organised to receive information from sensory inputs and use this to help us make sense of the world. As such they create a context in which we live our lives. Our senses are, therefore, the interface between the environment and the substrate that creates intelligence. If the inputs to the brain change, then neurones adjust and reposition their connections. This change in the inputs causes the substrate to restructure. Todd's question was whether adding a totally new sense could make his brain alter its structure and end up generating a new component to our view of the world. Ultimately, will human intelligence change as a result?

With his thinking partners, Todd therefore set out to create a whole new sense. Their ideal would have been an elegant, sophisticated and robust method of adding a totally a new sensory experience. Doing that would be more than complicated, because it would require them to attach the new receptor directly to a set of hundreds of nerve endings, and while people are finding ways of growing nerves in laboratories on plates that allow direct connection with nerves, developing ways of permanently integrating nerve-electrode junctions has proved deeply problematic.

Having gone through a range of ideas, they realised that a magnet was plausible with low risk. By implanting it in the fingertip it could piggyback on the highly developed pressure receptors at the end of each digit. There's no nerve connected directly to the magnet, but when Todd positions the magnet in the sort of oscillating magnetic field given out by electric motors or transformers, the magnet jiggles around. This movement is sensed by the finger's pressure sensors and reported to the brain. When Todd places his finger in a static magnetic field he can feel the magnet get pushed or pulled. To an extent it is not a whole new sensory experience, but it is a novel means of triggering an old one.

Not for the squeamish: pushing a magnet into your finger is not something to do without careful thought.

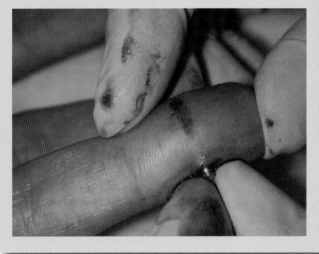

REPRINTED WITH PERMISSION FROM TODD HUFFMAN

Risks and Dangers

The implant went in simply, and Todd has still retained all of the normal sensation in his fingertip. That said, he is not naïve to the dangers involved in inserting an experimental implant and spent considerable effort minimising the risks. He is right handed, so the implant went in his left hand. The little finger was too small, and the index and middle do most work. His ring finger was big enough, and if worst came to the worst and it became infected and needed amputating, it would be least costly to lose.

The internal environment of the body is highly corrosive, and if neodymium corrodes it would create some nasty products. So before inserting his magnet, he and Steve built 10 implants and soaked them in warm salt water. One of the magnets was discarded after its imperfect casing allowed the saline to leak in and corrode the neodymium. The remainder were examined under a microscope to pick the one with the most uniform coating. At the same time they considered placement of the magnet carefully. In the end they implanted it off centre so that when Todd grabs something, the bone doesn't touch the magnet and puncture the coating.

'And several people have had implants go bad. I've seen pictures of the procedures where it's taken out, and it's pretty gross', commented Todd. 'When an implant goes bad, the finger just swells up and you have to surgically cut the magnet out.'

Todd's appreciation of the risks involved is one reason why he doesn't get involved in helping others put magnets in, though a few hundred people now have magnets positioned in fingers and arms. Another reason is that, like many, Todd believes that risks are only worth taking if there is some definite benefit. For Todd the benefit is the research drive to see if he can develop a new sense of reality, but he thinks that not many people will understand this drive. Consequently he believes that they would be left handling the risk without gaining from the benefits.

Feeling Magnetic Fields

So Todd can feel when he's around anything that creates a magnetic field. He can feel light fittings in display cabinets, or security detection systems in shop doorways. While it would be fascinating to feel or visualise the patterns made by the electromagnetic fields around us, Todd's implant is more of a probe which can be dipped into a field at a particular point. But does it have any practical value? Is he enhanced?

Well he can certainly sense things that others miss. When he uses his laptop, his magnet-finger rests over the hard drive. This storage device has stepping motors that whizz the read and record sensor over the surface of the spinning disc. If the computer's software is operating smoothly, the stepping motors have an easy time and aren't too active. If the software is asking the computer to access information from different parts of the disc, then the stepping motors are constantly active. If the computer has to wait while it moves its sensor to the best part of the disc, then it slows down, and if the arm is constantly active, then it uses more power and reduces the disc's life expectancy. Well-written software will therefore aim to minimise this effort. Todd believes that if programmers had his implant, they would be able to feel how efficient their work was, and that this could become a valuable additional tool enabling greater creativity. Others might be able to hear the relationship, or even in a partly assembled machine see it. But he can just feel it, all the time. For Todd it no longer requires abstract intellectual understanding, he just senses the interaction.

Todd can also tell whether an electric socket is live by simply plugging his laptop's transformer in and feeling whether the transformer starts working – he doesn't even need to check the little light on it; he once diagnosed a fault in his pond pump without stripping it because he could feel the motor working and thus concluded that it was the pump section that was at fault and not the motor. But for him it is none of these that make the implant useful.

'The magnet allows me to analyse my own thoughts about the world, because now I've had an alteration in my sensory input I can have a slightly outside perspective on what sensory input is and how it moulds thoughts', he commented. 'I consider that useful, and totally worth putting in implants. Most people don't think of that, when they ask whether it is *useful*.'

> ❝ The **magnet** allows me to analyse my own thoughts about the **world**, because **now** I've **had** an **alteration in my sensory input** ❞

Novel or Novelty?

Todd clearly has a capability that no other human beings had before him. You may question the scale of that capability, but it is definitely new. But would that make him a transhuman, and more importantly for me, is he enhanced?

'So what does it mean to be transhuman?' he replied, when I raised the issue. 'I find it more natural to think that humans were only human for probably only a really short period of time. There's this little slice of evolutionary time where humans were diverging, where they were kind of just standing naked in the woods. And then they started developing technology, and that enabled them to do things that a naked, unaltered human was unable to do. From that point on they were already heading down a transhuman path – they were early transhumans.'

To illustrate his point, Todd points to language. Talking and writing allows a person to transmit his or her information into another person's head. This information may convey ideas or your own private thoughts and they can reach out to someone else who's not even present. Aristotle, for example, continues to influence people because his written ideas are transmitted not only through space, but also across time. Although he lived in Greece a few thousand years ago, his ideas reach people today throughout the world. That, he claims, is not something that a naked human, standing by themselves in the woods, is able to do.

My problem with this line of thinking is that on this basis, everyone is a transhuman and so it devalues the value of the term. If we are all transhumans, then this is just another term for human. Pursuing the idea, however, leads Todd to an interesting observation. This transhuman quality, this ability to extend beyond biological boundaries, is largely due to the social nature of our being.

Take the internet. Now, having an isolated computer is OK, but it's vastly different when you connect a whole bunch of them in a network. There is an emergent phenomenon that is dependent on having lots of individuals communicating, co-operating and collaborating on things. Todd points out that we wouldn't have evolved if we hadn't been able to communicate, co-operate and collaborate – our lack of claws and big teeth leaves us vulnerable to attack and in a poor position to hunt. Our large brain would give us a chance, but the best use of it is in clubbing together and working as teams. It is when we work as a group that we extend beyond our individual human capabilities – that we extend beyond being simply human. This thinking leads Todd to claim that the actual human state only existed for an infinitely short point, and that in effect we've always been on a trans-humanist path.

I'm quite attracted by this line of argument, but then is this just playing with words? Is the true human purely a pre-Neanderthal, or is the very nature of being human to seek relationship and community, and to seek development? Is there any need to rebrand ourselves as transhuman?

So where does this get us in terms of the hype and hope of human enhancement, and where Todd's magnet fits in? I'm intrigued by the notion that enhancement only works in society, or at least takes on a whole new power when applied to social groups. Much of the argument and debate from people wanting to pursue enhancement is that they have a right to do what they want with their bodies and their lives – they want to exert their autonomy. What Todd is suggesting is that that autonomy only works when it is played out in a community of people who agree to work together, who effectively agree to relinquish some, at least, of their autonomy.

Modification and Social Responses

Traditionally, human body modification is a social thing; it's about group identification. The significance of alterations like lip plugs is not always obvious to those outside the group, but has deep significance within it. In Papua New Guinea, tribes decorate themselves elaborately for social events involving group dancing. Their adornment can include sticking feathers, bones or sticks through the central cartilage of the nose.

When a young Mursi girl in Ethiopia reaches the age of 15 or 16, her lower lip is pierced so she can wear a lip plate. The larger the lip plate she can tolerate, the greater her enhancement, the more cattle her bride price will bring for her father.

For them the body modification and decoration describes the state of relationships between members of the clans or smaller groups whose men all dance together. It announces their status, their social cohesion and teamwork. To procure and put on the costume is a sign of a healthy, loving clan, where there are no disagreements or feuds or doubts to poison relations and sap confidence. Anthropologists say that it's about looking the same, dancing as a team, in time, together. It's about the self disappearing and the society emerging. The sense of beauty and power is enhanced when you collectively look the same. A similar effect is created when riot police show up dressed in uniform and marching in step – you see the power.

It's **about** the self **disappearing** and the society emerging

I am struck by the way that, although Todd's motivation for his implants is an individual philosophical quest, he speaks of the decisions and the process of

A woman from the Mursi, a nomadic cattle-herding culture living in southwestern Ethiopia.

having the implant as a social thing, as a piece of teamwork. He didn't just go out and pay a design company and medical people to do this for him; he worked in a collective environment with a group of friends to achieve a technological feat in a cost-effective manner.

In parts of Papua New Guinea, tribal dress has many layers of significance. It is a visual language that conveys meaning and establishes relationships within and among communities.

Conformity and Visible Modifications

It turns out that there is more modification to Todd than his magnetic finger. Protruding from his right cheekbone is what at first sight looks like a small black dot 3 millimetres across, just under his eye, like a miniature beauty spot that escaped. In fact it's a small trans-dermal implant, a plate that is fitted to his cheekbone and protrudes through the skin. The post is machined to have a fine thread, enabling Todd to screw on decorations.

This gives Todd an unusual capability when it comes to wearing piercing-type jewellery, and Todd has turned wearing different items into a research project in itself. Although he is the first to admit that piercings are a lot about vanity, he is equally quick to claim that that vanity has a function – this sort of adventurous jewellery sends out powerful social cues. The companies Todd works for are in different states in the USA. He spends much of his life on aeroplanes. Before he had overt piercings he claims that his conversations with fellow travellers were dull in the extreme – limited to the weather. Now he tailors his decorations to his mood. The more extreme the earrings and attachments he wears, the more likelihood there is of causing other passengers to drop their guard and open up in conversation. The more extreme his fixtures, the more likely people are to open up with weird stuff about themselves. Conversely, Todd has more subdued jewellery for the moments when he enters boardrooms.

Todd, the pink Viking in the centre, loves an excuse to get into costume.

So here's a man who chooses to conform to one peer group at the time, yet who claims not to like conformists. He took the risk of having a magnet implanted inside his finger when no one else had done this before, to explore a philosophical question. Yet he uses a second implant to amplify the social cues his clothing give and he's sensitive to the shared language such cues operate with.

He's not alone in seeking to modify his appearance to fit in. Plastic surgery offers you the hope of looking like the person you want to be. Choosing the clothes you wear has a similar effect. You wear a suit to go to a memorial service, and indicate your sense of mourning, you wear a bow tie to a black tie dinner – unless you deliberately want to make a non-conformist statement.

Unless it has an ornament attached, Todd's modification is much more difficult to see, but it does place him firmly at the centre of a very active international, social and philosophical group. Indeed, his modification has led to him having an enhanced status within his chosen community.

To Be, or to Be Made?

'So transhumanists as a group are generally believers of Progress, with a capital "P". We have this trend of using technologies to change the human experience. Transhumanists usually believe that technology can be used to make the human experience more positive and better', Todd commented.

There is an interesting comparison with observations of body shaping and lenhancement in many ancient civilisations. Take the Panará peoples of Brazil as an example. Here, the parents manipulate babies to give them more desirable characteristics. They stretch out and stroke babies' legs, because long legs are regarded as elegant. They tie string around the top of the calf to make the calf muscles very rounded, and they massage the heads because they believe that a round head is good. For them, growing up will not happen automatically. It is the result of the parental action. They see it less as a process of altering humans, more one of making them. For these people, modifying their bodies is not about improving their experience as humans, it's about ensuring that they *become* human.

To my mind, Todd's view of transhumanism and human enhancement seems to be playing a similar game. The natural human is something that no one wants to see wandering the streets of any modern city. Any child needs to be quickly modified through culture and education so that they can cope with the developed society we have built, and then encouraged to push for more change, in both themselves and their society.

'I'm a huge fan of the individual, but even the glorious individual is only there by an accident of history; where you're born, your schools, the society and the environment around you – you've no control over where you enter the historical stream', Todd added. His point is that in order to achieve large, complex goals, like those that transhumanists seek, you have to include other people. No one on their own could upload themselves into a supercomputer. You would need other people to help. The marshalling of societies brings together people with different skills and abilities. You may not know who they are, but together their achievements can be vast. History has clearly shown that by acting together we are capable of improving the human condition.

The Importance of Being Implanted

You will remember that Peter Houghton felt he had much in common with people who had external ventricular assist devices. He thought it made little difference if the technology was internal or external if you were equally dependant on it. Internalising it gave some practical advantages but didn't really change how much a part of him it was. For him an enhancement could be external – the internal/external distinction can't be used to define an enhancement.

For Todd, however, it is important that the implant is internal. When people ask why he didn't just tape a magnet to his finger (and, he says, they frequently do) he responds with two levels of answer – one practical and one philosophical. Practically, implanting the magnet greatly increases the fingers sensitivity to its tiny movements. Without that high-fidelity feel, the magnet would have little value. Philosophically, he also believes that there is a big difference between bolting on an external artefact, and having something that is built in, that is part of yourself. It is also important to him that the implant is intended to be there long-term; it's not just fixed for a thrill and then thrown away after a few weeks.

One way of getting the idea is to think of someone who's blind. You can explain the physics of sight and describe how the eye works. You could even create string maps that indicate how neurones are connected in networks. All that, however, would not let you explain the experience of sight to someone who has never seen. That internal experience only comes when the sense is part of you. 'Temporarily taping a magnet to your finger would make it an external and not part of you. Your mind won't internalise it as a natural phenomenon', says Todd.

But why can't you view an external thing as part of yourself? Pete Houghton is dependent on his external battery pack. I've worn glasses since the age of six. Each morning they are one of the first things I reach for, and each night they are among the last things to get removed. They are external, but distinctly part of me. There are few pictures of me without them and for most people they are part of my identity; without them my sensory perception of the world – my ability to see – would be radically worse. But they are external. Some children only develop their sense of sight because they are given glasses before they even make it to school – in fact I'd probably not have sight in one eye if I'd not had glasses as a child.

Todd is convinced that having the magnet inside his finger makes a difference. He told me that over the first few days that the magnet was in place he went around feeling for fields. The time he started to learn new things was when he stopped looking for sensations, and got on with living. Then magnetic fields would catch him by surprise and trigger his interest. He feels that the permanence of an implant made it integrate with his life.

Maybe I don't see it his way because my fingers don't respond to magnets – but I think it is the permanence of use, rather than the internalisation, that is important.

Therapy and Enhancement

One recurring question has been whether an intervention is enhancement or therapy. Todd brings a slightly different slant to the issue, because for him enhancement will only move forward in a society that has a highly developed therapy-providing healthcare system. The result is that while he enjoys his magnetic-enabled finger, he feels that the main focus for enhancement needs to be dragging healthcare into the technology age.

We are back again to the idea that enhancements are things that happen to individuals when they live in enhanced or enhancing groups. Just as Aubrey de Gray is working to develop therapies that tackle all routes to ageing, Todd is looking for a society that has the health resources that will tackle disease and damage. That way we can start to build a better future.

Todd points to laser eye surgery as an example. Some competitors in sports who require finely tuned vision, such as shooting, are turning to laser surgery to perfect the surface of their eye and create a better lens. This has only been possible because laser surgery has been developed as a therapy for people with poor vision. The aim there was not to create super vision, but to pull a person with visual disabilities towards normal. Once the technique is up and running, however, it is not difficult to extend it. 'You're not going to build enhancements without a base of therapy', comments Todd.

Enhance to Conform

For Todd, enhancing is conforming. Modification is about making a better future, about continuing human beings' transhuman trajectory. In many ways his social approach to body modification and enhancement is more like that of traditional tribes than it is akin to the individualist Western fashion and popular culture of body art.

Modification is about **making** a better future, about **continuing** human beings' transhuman trajectory

But could Todd's philosophy be extended to force people to become more transhuman? Could it lead to a situation where unmodified, unenhanced people are less valued, are less human, in the same way that the Panará Indians do not give moral status to unmodified children?

It strikes me that this is a real possibility, but it is going to take something vastly greater than a fingertip magnet to trigger that change, and the vastly greater implant appears to have eluded me so far.

8 Adding Technology

Kevin Warwick is a man who enjoys making headlines and takes grabbing the front pages of tech magazines as marks of success. A pivotal moment came when he filled the front page of *Wired* in February 2000 (http://www.wired.com/wired/coverbrowser/2000). His claim was that he was now Cyborg 1.0, the first in what he claimed would become a line of increasingly advanced human-technology amalgams. The article started with a statement from Kevin: 'I was born human. But this was an accident of fate – a condition merely of time and place. I believe it's something we have the power to change.' In the heady millennium year this went down particularly well, as every media outlet was searching for the most extreme comments on our future trajectory as a society and a species.

> **I was born human.** But this was an accident of fate – a condition **merely** of **time and place.** I **believe** it's **something** we have the power to change

His love of publicity has made Kevin as many friends as enemies. Media folk love academics who are prepared to put their heads above the parapet of their ivory tower and make exciting statements in normal English. Most academics tend to do the opposite. They sit tight, hunkered down in their labs and making forays to conferences where they talk in jargon with the small number of people who understand the code. They are deeply suspicious of any move to expand their audience by addressing the media, because they fear that the true depth of their wisdom will get lost in the lights.

As an academic turned journalist and author, I recognise both sides of this issue. I was once was cited in *The Times* as the inventor of the artificial placenta – a feat that has not been achieved, least of all by me. For a few days it caused great embarrassment to me and close colleagues, but in the great scheme of events, the error did not change the world. I now stand on the other side of the divide, trying to get scientists to explain what they are doing, very often in jobs funded by the public purse. Their reluctance to talk is often justified with a phrase like 'I don't want to become a Warwick'.

I wanted to meet Kevin for myself. He is someone who has implanted devices into his body with a desire to give humans new capabilities, and it struck me that in my search for understanding about enhancement, he would be a useful foil against which to check the balance between hype and hope.

Defining Terms

We met in his office in the Cybernetics Department at the University of Reading. It's a classic academic's workspace with floor-to-ceiling shelves groaning with books, and extra piles of books and papers on every other horizontal surface. A few electronic gizmos dotted around suggest a hobby-like interest in circuits.

To be an enhancement it would also have to be integral, it's not like wearing a pair of glasses or driving a car

Like most people involved in enhancement, Kevin was less than direct in answering the question, 'What is enhancement?' Like all the other enhancement enthusiasts I'd met so far, he rehearsed the problem of agreeing on a definition and pointed to the different options. Then he reached his version. For Kevin, enhancement involves giving abilities beyond those that we would normally consider a human to have. To be an enhancement it would also have to be integral, it's not like wearing a pair of glasses or driving a car where the devices are external to you.

Not surprisingly, Kevin therefore thinks that Todd Hoffman's magnet is a good example of an enhancement, because it is a new ability that is integral to his body. This line of thinking would, however, rule out Peter Houghton and his mechanical heart pump, because it is not better than the original, does little if anything new, and has part of it on the outside of the body – it's an analysis that would meet Peter's approval.

One advantage of Kevin's definition is that it makes an attempt to draw a clear distinction between enhancement and therapy. The desire is to step beyond the norm. The problem however, is that we are back to the issue of trying to define 'norm'. It's very difficult to know what powers and abilities some of the extreme members of the human race naturally have – many after all can run 100 metres in around 10 seconds. Give it a go, and you will see that that is a fairly extraordinary feat – but still something that some unenhanced, though trained, humans can perform. At the same time, when these sorts of enhancements are attempted, they generate considerable media interest with consequently the distinct potential for generating hype.

Kevin doesn't dispute that the other ideas are interesting, but wants to focus on what he admits are the more sci-fi type of enhancement. He wants to focus on super powerful legs and the six million dollar man type of physical and mental enhancement, although he is quick to acknowledge that it will never be easy to draw a neat border between enhancement and therapy.

Kevin Warwick's Career and Enhancement Timeline

1954   Born in Coventry, England
1970   Left school at 16 and worked for British Telecom
1976   Went to Aston University, and then Imperial College London
1987   Became Professor of Cybernetics at the University of Reading
1998   Implanted radio transmitter in his upper arm
2002   Implanted 100-pin electrode in nerve in lower arm

The Cat-Flap Chip

The first time that Kevin tried to increase his human abilities was in 1998 when he had a small 'chip' inserted into his arm. It was the size of an extended grain of rice – about two millimetres wide and 15 millimetres long. At one end was a couple of very small computer chips and at the other, a copper coil. The chips ran a small radio that transmitted an identifying code, and the coil acted both as an aerial and as a means of picking up power by electromagnetic induction.

The modification to Kevin was small and short lived – he only had it in for nine days – but to make use of it required considerable modification of his environment. Doors in the office area where he worked were fitted with transponders that beamed out energy and waited to pick up the identifying feedback from his chip. If the signal was there, the door would swing open. Other receivers turned lights on as he walked into a room, or triggered a computer to play the greeting 'Good morning Professor Warwick!'

So was the microchip-implanted Kevin an enhanced human? He certainly claims so because he says he felt a special link with the world around. He liked doors opening and believes that other members of staff began to feel jealous of his new powers. On top of that he just found it 'good fun'. I wonder whether it would have been just as fun if the technology had gone wrong and slammed doors in his face, giving him a bad house day!

Kevin's chip – you can see the original in the Science Museum in London.

He liked doors **opening** and believes that **other members** of staff began to feel jealous of his **new** powers

I struggled to see why implanting it had any more power than just slipping it in his pocket. Kevin's response was that having it under his skin made him feel as if it really was part of him. He equated it to the implants placed in Parkinson's patients of heart pacemakers, saying that mentally they feel that the thing is them, and people who have received transplanted organs soon regard the new body part as being them.

Implanting the device did not require major surgery, but did raise significant issues. Did it also lead to enhancement?

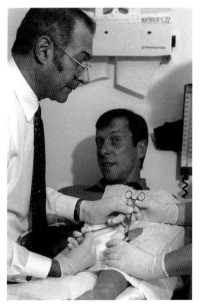

My concern here is also that Kevin is making great claims for a small bit of very simple technology. The implant was little different from the transponder that comes inside fobs you attach to cat collars that let only your cat open the flap, and I was a little disconcerted when Kevin agreed that it was a glorified cat flap. The power of the experiment for Kevin was in seeing whether implanting this technology would have any negative effects. To emphasise this, he called the experiment 'Big Brother'. Wherever he walked in the building the computer tracked him. It recorded the time he entered each room and the time he left. People who had access to the computer knew where he was.

'But I never regarded it as a negative. It gave me a sense of power and I adapted pretty quickly to not having to open doors or look for light switches. In fact, I was quite disappointed when it got taken out', he said.

But, I wonder, would Kevin's enjoyment of the implant have changed if he hadn't had access to the computer that was tracking him, if someone else was watching, but he had no control over it? A Big Brother question would involve not only being watched, but being controlled, being kept in order. How would he have felt if the doors had locked if he showed up two minutes late for a meeting and had to stand outside for a further two minutes as punishment? What would he have done if his salary had been docked for leaving early or showing up late? Would he have liked it not only if his current whereabouts were known, but if exactly who he was with at each moment of the day was open to public scrutiny? I'm sure that would worry some people. Indeed it didn't take long before Kevin could see that less benign ownership of the data generated could soon lead to the experience being less than pleasant.

It's a **bit** like extolling the virtues of **explosives** by showing how **useful** they **have been** in mining, without **investigating** their role in bank thefts and **warfare**

You can see why Kevin is an easy target for criticism. It strikes me that he called the project 'Big Brother' because it would grab headlines. But he was then unwilling to introduce any mildly Orwellian features to the protocol, so was almost bound to come out thinking that the chip was a wonderful toy. It's a bit like extolling the virtues of explosives by showing how useful they have been in mining, without investigating their role in bank thefts and warfare.

Kevin pointed to the fact that we live in a world where people increasingly know vast amounts about us. Every time you use a credit card, the authorities can track down where you are, so the negative aspects, he contends, are not new. 'We are prepared to put up with that because we like the benefits that come from not having to wander around with cash', he commented.

If Big Brother was a somewhat hyperbolic term, how about his claim that this chip made him Cyborg 1.0?

'I think calling me a Cyborg was as much to make people think and ask questions. Rather than saying here's the cyborg, I wanted people to ask "Is this a cyborg?' Is it a big deal to have it implanted or not?"

Since then the idea of adding identifying chips to people has cropped up in a number of different environments. One chain of clubs have begun to offer clients the possibility of having a chip inserted in their arms, instead of carrying around a membership card. This way you become your card, and the photo on the card is replaced by your face. The Baja Beach Club in Barcelona or Rotterdam, for example, offers the choice of card or chip. It makes for interesting publicity, but once you have the chip you don't need cash for drinks: wave your arm and the bill is on your account. People with diabetes and epilepsy can also have chip implants that record vital information about themselves and their condition. It's a high-tech version of a medical emergency bracelet.

In Mexico, the technology-aware government has started chipping its senior staff. In November 2004 they placed microchips in 160 people who work in the Attorney General's office. Without these they were unable to gain access to restricted areas. In many ways this is no more than the countless security badges that many people wear as they wander around their offices and laboratories. But the fact that it is implanted does also mean that you can't go home, take off your badge and go out anonymously. Does that make you enhanced, or cornered?

Each time a missing child hits the headlines, Kevin says his phone starts ringing with people interested in placing chips in children. In the summer of 2007 it was Madeleine McCann in Portugal, in 2002 it was Holly and Jessica.

'Right now we have no idea what happened to Madeleine, but she's not with her parents and it looks as if she's been taken without their knowledge or consent. With an implant she would have been found, you know, within 24 hours for good or bad but that would have been it', said Kevin.

But we need to stop for a moment and check my hypeoscope. If these chips are going to have any hope of sorting out this type of missing child situation, they would either need to be much, much more powerful so that they can 'shout' to remote sensors, or we would need to build sensors into every public place and link them to a central computer. Both of those situations are outside the scope of current technology. The chip would need to act like a mobile phone, possibly even a satellite phone and would demand too much power to run for more than a few minutes at a time, unless it was supported by some form of external power source. And the genuine Big Brother fears of a government agency tracing all people-movements through all public doors would meet with considerable resistance, even if it was technically and financially feasible.

Enhancing Groups

Implementing this type of technology is much more than letting individual parents fit their children with a device. Being the only one with a device is less than useful. Take something like the telephone. When it was first invented it looked like a gimmick, a toy. It was difficult to see how it could do anything useful. The benefits came when lots of people took up the technology. Likewise, Kevin admits that the first chip-enhanced people are also gimmicks, but believes that that will change if many start to adopt the idea.

He complains that most people are used to looking at the world as it is now. They do a pretty poor job of thinking outside the box and looking at what an enhanced world could be like. For him, the challenge is to get away from present-day understandings of ourselves and our communities and think ahead. I agree with the challenge, but I'm also keen to keep an edge of realism when out horizon-scanning.

Kevin wonders what an army would look like if each person were fitted with a transponding chip. Would you be able to avoid friendly fire incidents? Again, however, I wonder how far you can go with this. Any technology that identifies a person as friendly to their own forces has the ability to be deciphered and mark them as foe to the 'enemy'. It could easily be the technological equivalent of painting a target on your back.

It **could easily** be the **technological equivalent** of **painting** a target on your back

Collective Consciousness

It's relatively simple to detect somebody's level of excitement by monitoring how easily electricity passes across their skin. One idea is to give clubbers a little device that measures this feature and has a Bluetooth link to the DJ. By analysing the overall level of the dancers, the DJ could then adjust the pace and intensity of the music and constantly monitor the response. A consequence would be that the music would start to evolve under the collective response of all the dancing people.

If you built an implantable device, maybe it could monitor sugar levels and hormone responses, giving even more information to the mood machine.

Joining in would of course involve the risk that those in charge could change the mood of the group against their will. You would need to be either high on trust or low on caution.

Identity in Community

Let's pause for a moment and think. The current trend in discussing the implementation of technology is to argue for individual autonomy. The theory is that people should have control over their own lives and bodies – to be autonomous decision makers. The problem is that your decision to have a chip-enhancement is only going to work if you can get a large chunk of society to go along with you. In addition, your desire to be enhanced by placing a chip in your arm will get nowhere unless society builds the rest of the kit that makes the implant come to life.

Again it seems to be society rather than the individual that is enhanced. Your choice then is whether to join in with that enhanced society. This idea of an enhanced society is the subject of a European research project that Kevin is taking part in. The idea is that around 50 people will be fitted with tracking devices for a month, allowing anyone who is interested to track them moment by moment. 'It's a social science experiment. You will be able to log on and say "Ah so Kevin Warwick is there, this person is there, this person is there" ', he said.

Personally I'm not sure what his home insurance company will make of him broadcasting to the world each time his home is empty . . . because again, information gathering is fun, so long as you can restrict who has access to it.

The Second Implant

The implant that Kevin played with in 1998 was technically pretty simple, so in 2002 he tried again, this time with a more complex device. Instead of simply being slipped under the skin, like a silicone-coated splinter, this implant plugged into one of his nerves.

Instead of **simply being** slipped under the skin, like a **silicone-coated splinter**, this implant **plugged into** one of **his nerves**

The active head of the implant was a small plate containing 100 1.5-millimetre-long pins. The whole plate was only three millimetres square. In a tricky operation a surgeon exposed the four-millimetre-wide median nerve, which runs near to the skin just above the wrist, and chopped a small square window out of its protective sheath. He then fired the probe into this hole, so that the pins penetrated the bundle, and many of the pins came close to individual nerve fibres.

The second of Kevin's implants.

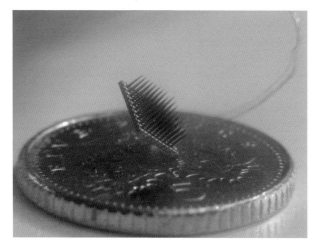

The surgeon tunnelled a small cable carrying tiny wires connected to each of the 100 pins up his arm, and brought it to the surface through a small incision. At this end the cable was fitted with a connector that let the device be connected to a computer. The researchers could then either try to pick up signals from the nerve, or transmit signals down into the nerve. Even though there were 100 pins in the device, the technology limited them to addressing only 25 at any one time. He had it in for just over three months, and then took it out. One hundred days is a reasonable time for performing experiments, but it is worth noting that the implant was beginning to lose function by the time it was removed.

Kevin and his team embarked on three sets of experiments. One was to see if he could build into his body a new mode of sensing the world. His choice was ultrasonic 'vision'. To do this he built ultrasonic sensors, much like the ones currently fitted to the rear bumpers of cars, to either side of a baseball cap. As the sensors approached an object they sent a buzzing signal into Kevin's nerve,

Inserting the second device required more serious surgery, and the risks of permanent damage were greater.

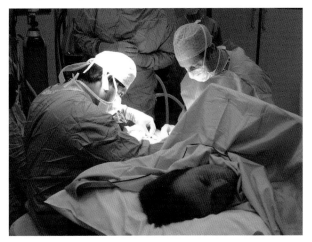

sending the signal from the left side sonar to one set of pins, and the right side sonar to a second set. With his eyes covered, he tried to navigate around the lab.

As the **sensors approached an object** they sent a buzzing signal into **Kevin's** nerve

Within six weeks he could recognise how far objects were away and move around without bumping into things. He could also follow a large board that one of his researchers held in the air. For Kevin, the interesting thing was how quickly his nerves learned to listen out for a novel signal, and his brain learnt to let him interpret this in terms of the location of an object. The actual practical utility value of being able to follow a board around a lab was secondary to the fact that the brain was not such an old dog, and that it could learn a new trick.

With sonar on his cap linked to the electrode implanted in his arm, Kevin could follow a board while blindfolded.

Risk of Implant

About three days before the operation Peter Teddy, the neurosurgeon, took Kevin to one side to check that he really understood the risks. They included:

Infection – the operation could introduce bacteria into the arm, and the hole that the wires ran through would give a permanent access point for bacteria. If an infection had got in at the site of the computer plug, it would run very quickly down the moisture that lined the cable and reach his nerve, where it could do significant damage.

Nerve damage – removing the nerve sheath is a tricky task and would leave an area of permanent weakness. Firing the electrode into the nerve could also harm the delicate nerve fibres. This could result in a loss of sensation in the hand, as well as a loss of control of hand and fingers.

'It's not too late to say no', the surgeon had said.

To show me how easy it is to make sense of a sonic input he got me to put on a similar baseball cap. In this case the two sonars were linked to vibrating cylinders that I held in each hand. I closed my eyes and attempted to walk around the office. It took only a few second before I could gauge where high filing cabinets and walls were. It seemed simple to use, but apparently there are about one in 10 people who just can't get it. I got a little more adventurous and reversed, almost instantly tripping over a chair that was at knee level. Being six foot three meant that the head-mounted sonars had no chance of warning me about low-lying obstacles.

I got a little more adventurous and reversed, almost instantly tripping over a chair that was at knee level

'The point is, you've just put the hat on and clearly your brain has made the link straightaway, these signals actually mean something – it's able to do the translation from buzzy signal to a perception of what is around. The difference for me was that the buzzy vibrations were effectively in my brain', explained Kevin.

The research team had an interesting time trying to decide what current to use to stimulate Kevin's nerves. The challenge was to find a signal that was strong enough to stimulate them, but not so strong that it fried them. This was made all the harder to start off with, because when at first Kevin couldn't sense anything, they didn't know whether this was because the signal was not strong enough to stimulate the nerve or the brain was not recognising the inputs. After a few days, however, Kevin could detect a signal, and over the following days he became more sensitive to the input. 'Presumably my brain became better at recognising it over a period of time', he said.

An ultrasonic detector works by sending out a pulse of sound that is at too high a frequency for the human ear to hear – it is ultrasound. It then waits to see how long it takes before the sound bounces off an object and comes back again. The longer it takes for this echo to arrive, the further away the object is.

Despite its apparent success, since 2002 no one else has tried a similar experiment. It's frequently referred to in other papers, but in reality this work is in a state of limbo with no new work happening. Kevin, however, would love it to be used by people like firefighters who would then be equipped with a coat of many sonars, and capable of working in a smoke-filled environments because they could 'see' through the smoke. The thought of this sets my hypeoscope squeaking again. Vision works because our eyes have six million rods and around 125 million cones, each of which sits there detecting incoming photons and creating a detailed picture. It can have so many detectors because the wavelength of light is very short – or at least very short in comparison to sound waves. Even if you could cover a jacket in sonars, you'd never be able to pack in enough to get anywhere near the same sort of definition.

That sort of analysis doesn't dampen Kevin's enthusiasm for the experiment. Even a rudimentary perception driven by sonar fits with his idea of enhancement – it's certainly not something built in to normal babies, though he admitted that most of the technology and the power supply was outside him – it was only the nerve interface that was internal.

'But just as we ask what is it like for someone who has never seen to suddenly be given vision, or someone who has never heard to have a cochlear implant and hear, we can now start to ask what it would be like to live with an ultrasonic sense?

We can start to take seriously the idea of having enhanced sensations. Was there a direct benefit from my sonar? I doubt it – but it raises the issue to a higher level of debate', Kevin claimed.

Sending Signals

A second set of experiments with the implanted nerve pins was to see what signals could be picked up from the nerve when he moved his hand, and see whether they could use these to drive prosthetic devices, or even develop novel means of communication.

A couple of weeks after the operation, the team was able to record signals from around 20 of the 100 pins that had been fired into his nerve. Kevin could move his hand around and the computer recorded clear bursts of nerve activity. The exciting thing for him was that different movements created different sets of activity between the pins. The reason for this was that each pin sat next to a different combination of the 10,000 or so nerve fibres, each of which heads off to control a different part of the muscles in his wrist and hand. Altering the movement altered the spectrum of muscles needed, and so created a fresh set of recordings.

Once something is implanted in the body, the immune system recognises it as foreign. If it can't get rid of it, the immune system sets to work isolating it by building a sheath of tough material around it. For Kevin this was helpful in that the sheath would hold the implant in place, and it would also help to isolate it from the rest of the electrical activity in the body. And that isolation made it easier to record nerve activity.

'How much of the complexity in the nerve signal could you distinguish?' I asked, recognising that when you record nerve activity you look at the difference in electrical activity between two pins, so having 20 working pins gives a large combination of possibilities.

'Well we could pull just about everything out of it; actually making sense of what there was was a difficult thing', he replied.

The hypeoscope beeped again. There is a phenomenal amount of activity going on in 10,000 neurones – to get 1% would still have been a massive result. His claim to be able to record 100% worried me, because it suggested that Kevin either had slipped in an exaggeration or didn't know his nerve science all that well – possibly a combination of both.

Still the information they pulled out had some power. After a bit of training, Kevin could wire a robotic hand into the computer and when he opened and closed his hand, the artificial fingers followed suit. On another occasion he linked the computer to an electric wheelchair borrowed from Stoke Mandeville hospital which houses the National Spinal Injuries Unit. By making movements of his hand he could control the chair, even though he never touched its joystick.

Always keen on a stunt, Kevin flew to America and plugged his nerve recorder into a computer that was linked to the internet. His robotic hand was left in

Reading attached to another web-linked computer. Guess what? When he moved his hand, the robotic hand moved. It's a great media moment, but in reality it says little more than that the internet can be used rather like a remarkably long extension lead. After all, we can do that sort of thing already. Commands entered in one place can be executed in another.

'Sensory input was sent back across the internet and I could feel how much force the hand was applying', Kevin explained. 'There was not much dexterity, but just a simple how much force it was applying. Suddenly your brain can control things on a different continent or maybe a different planet. Was is it better – was I enhanced? I don't know, it was just very different.' Kevin's experiment raises the possibility of using similar implants to drive electric wheelchairs for amputees or operate dangerous equipment without touching anything. That would certainly be outside normal ability – and would class as an enhancement. The technology though has a long way to go before it is ready to be installed.

On a US visit, Kevin watches a prosthetic hand move in Reading, UK, via his web-linked implant.

Suddenly your brain can control things on **a**
different continent or maybe a **different** planet

It's Good to Buzz

Kevin's third experiment involved trying to enhance communication. For this, the team placed a simpler electrode inside one of the nerves in his wife's arm. When Kevin moved his hand, the signal that travelled in his nerve was transmitted across to her nervous system. 'We had a telegraphic form of communication

Kevin and his wife share nerve-to-nerve communication for a few hours.

directly between nervous systems', Kevin claimed, saying that for him communicating electrically directly from brain to brain was an incredible enhancement. 'So ultimately it will mean that we don't need this stupid form of speech to communicate. Is that benefit?' he asked, clearly implying that for him it is.

❮❮ We had a **telegraphic form** of **communication** directly between **nervous** systems ❯❯

What Kevin dreams of now is cutting out the middleman. Why use the limited signals that are sent down peripheral nerves and connect them together? Why not link brains to brains? Working with surgeons in Coventry, he is determining where in the brain to place an electrode so that it could both collect signals that could be sent to another person, and receive signals sent by that other person.

Enhancement/Benefit Balance

So once again I found myself trying to work out how much of Kevin's work is enhancement, and how much is standard scientific enquiry that could lead to greater knowledge and potential future therapies. And maybe that's what enhancement is all about. Maybe it's about what is just beyond our knowledge and capabilities now, and is what we still have to dream about.

9 Better and Beyond

I Skyped a lady called Fenella, an interesting way of communicating face to face with someone who is not in the room. In this case, she was only a hundred miles away, but she could easily have been anywhere in the world. Still, from my point of view she was in my office and we could see and hear each other perfectly.

Once we had got our webcams and microphones sorted out, Fenella outlined her story.

'I was born profoundly deaf. It's a one-off thing, no one else in the family was deaf and it just happened to me. My parents put me on hearing aids from the age of 15 months and as I grew up, they sent me to normal hearing schools. Obviously I had a bit of difficulty following the teachers, but my parents felt that I needed to learn to speak orally, and to integrate with the hearing world, for which I am very grateful.'

About four years ago, when Fenella was expecting her third child, she picked up an ear infection while swimming in the Mediterranean. It left her with almost no hearing in either ear. The reduction was such that wearing a hearing aid made no difference. She also lost her sense of balance and was very ill.

So Much More to Hear

Having been checked out by the Ear, Nose and Throat consultant, Fenella decided to have a cochlear implant. Fenella showed me an email that she had sent around the family describing what happened next.

From: Fenella
To: Family
Sent: Tuesday, August 23, 2005 8:13 AM
Subject: Cochlear Diary! 23rd August 2005

I thought that you might be interested to hear of my progress . . . I am trying to keep a diary of this important and wonderful phase!

I came out from Southmead Hospital in Bristol after having my long awaited for COCHLEAR IMPLANT which was on 23rd July, to the dreadful traffic with all the caravans making their way down here. Its lovely to be at home to be able to sleep for the first time properly in my own bed with my lovely soft pillows as opposed to the hospital starchy and springy ones! You only notice this when you have a sore head!!!!

The operation went well and was quite an amazing experience with the anaesthetic – totally wipes you out!! I remember all the stress and weepy episodes prior to the operation and then having the anaesthetic pumped in. On waking up I first said "right, I am now ready for theatre!!" and then one of the anaesthetic staff said, "Sorry madam but you have had it done!!" – what a lovely knock out it was!

The staff at the ward were super and kept saying that I was known as the "million dollar baby" giving me the initial impression that I was all knocked out and bloodied up like the female boxer in a film 'million dollar baby'. . . . but actually it was a reference to the sophisticated chip that I have had implanted into the mastoid bone and skull, which is stronger than the space shuttle.

I then had my 'switch-on' whereby the outer pieces of equipment were fitted to my ear with a little circle bit attaching itself to the magnetised chip planted in my

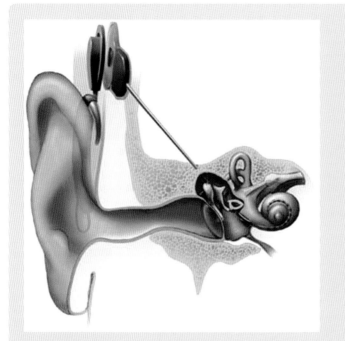

A cochlear implant is a small electronic device that is built in two parts. One bit sits on the outside of the body, behind the person's ear, and contains a microphone and a clever bit of technical wizardry that selects and sorts out the sounds, before creating a signal that can be transmitted to the second part. This piece consists of a receiver that picks up the signal and converts it to electrical impulses. These are fed down a wire to a series of electrodes placed on different regions of the auditory nerve.

The way it works is very different from a conventional hearing aid. Instead of simply boosting the sound and blasting it at the eardrum, cochlear implants generate signals that are sent straight to the auditory nerve. They therefore bypass the physical mechanisms that pick up sound in normally hearing people, and therefore leapfrog many of the problems in people who are deaf.

Hearing with the implant is very different from normal hearing and a person has to learn the task. The brain, however, shows a remarkable ability to adapt to this new input and soon a person is able to enjoy conversations and even use telephones. Research shows that the brain is particularly capable of changing before the age of two – the implants are therefore particularly powerful if put into children very early in life.

head. The first switch-on was done on 16th August and took 2 hours to do was really disappointing as things were so quiet even though I could hear Mike and my mother who were with me, but were very quiet (in Sound!)

Day Two, I couldn't hear anything . . . I was then increased up to a preset decibel level of 40. This is remarkable as before the implant I couldn't hear anything quieter than 140 decibels (Pneumatic Drill/Aeroplane). The average human ear is set to 20 decibels . . . so its not long to go before hopefully I shall be tuned into a normal hearing level.

The main thing is not only do I need to learn to hear the environment sounds but I will need intensive 'hearing' therapy to learn to listen to speech! This is anticipated to take no more than 5 weeks so I will be up to Bristol on a weekly basis to be 'reprogrammed'!

On this second day the pitches had to be turned up and it was then that I heard my mother who came with me and the audiologist's voices and my own voice which sounded even more high pitched! I heard pens tapping on the table really clearly . . . voices with lots of sssssssssss sounds . . . chair used to be 'air' became 'ccccchhhhhhhhhaaaaiiiiirrrrr'!!!! Really but but wonderful as the sounds around me became so Clear and crisp it really was wonderful! So many new noises I picked up, the doors shutting, my flip flops clopping away on the hospital floors, trolleys rattling in the background, several 'pinging' noises . . . and a lot of metallic noises!

Another day was a bit of an experience trying to cope with new high pitch noises which seem to be taking ages to get used to! I have noticed though instead of feeling tired from all the noise I end up feeling more sick! The sickness is like having too much Vitamin A if you have been out in the sun but its not that – it's the tiredness of my auditory nerve having to cope with all these new noises!!!

I have discovered more new noises such as kettle boiling, bread baker machine, a HUGE dragonfly humming when it came into the kitchen and couldn't get out again! Dilly's dolly crying, children shouting in the next room when I am in the kitchen! Umm, the water gushing along the hosepipe when I am going to water the outdoor plants! Various bits of clattering in the kitchen! Also, hearing the telephone ring from a distance, my mobile phone ringing,,,,

Perhaps I should keep a diary of all these new sounds each day! I do have a checklist to go through to see if I can listen and hear these noises! Also, I could hear the frying of eggs in a frying pan, the baked beans and rhubarb bumbling in the pans whilst cooking!!!

I still have another 4 more sessions at the hospital to go which hopefully will finalize the implant! It's amazing and absolutely brilliant!!

So many **new noises** I picked **up,** the doors **shutting,** my flip flops clopping **away** on the **hospital floors,** trolleys rattling in the background

▌▌ I was known as the "million dollar baby" . . . a reference to the **sophisticated** chip that I have had implanted into the **mastoid bone** and skull, which is **stronger** than the **space shuttle ▌▌**

A few weeks later Fenella found herself back at the keyboard.

From: Fenella
To: Family
Sent: Sunday, October 02, 2005 7:52 AM
Subject: Cochlear Implant Update!

I thought that now it's 1st October I should write more on the sounds that I am continuing to pick up through my cochlear implant and that there are so many it's difficult to try and pinpoint each one and write it down!

Hearing telephone ring from the next room
Hearing the computer battery life running out from two rooms away
Hearing the mobile telephone from the next room when a message arrives
Children having a fight in the next room!
Husband shutting and opening door from 3 rooms away
Dishwasher being loaded and unloaded from next room
Birds Tweeting through windows / when outside / more clearly
Quiet dishwasher pumping water throughout cycle!
Tumble Dryer going
Washing machine going
Shoes sliding across wooden floors
Water running through waste pipes
Lawns being mowed

House Alarm going off in Distance (ie 1 mile away!)

Several electronic pings during shopping expeditions!

Chairs scraping on wooden floors (likened to a chalk on blackboard – very irritating!)

Buzzards and Rooks making their 'territorial' calls

Male voices (picking up the deepness of vocal chords)

The waves crashing across the beach

Children sneezing (without looking at them)

Footsteps echoing across floors

Other people in a restaurant 'clicking' their knives and forks!!!

Hearing guests talk during lunch party REALLY clearly even though sitting a few feet away! Wonderful! and without having to ask guests to repeat or slow down their conversation

Keys clicking against one another during car drive

Hearing even more Nursery Rhymes eg. tumble tots 'come on everybody . . .'

'Old MacDonald had a farm' 'Hickory Dickory Dock' 'Baa Baa Black Sheep'

'Two little Dickie Birds sitting on a wall'

Some Christmas Carols on CD such as 'We three kings of Orient are . . .'

. . . .

. . . . .

I will try and update this more often and write down the most amazing sounds!

The most amazing sound, apart from her children, was hearing bubbles in a gin and tonic

Her experience is shared by the majority of more than 100,000 people worldwide who have had cochlear implants, and two years later Fenella is still excited by the sounds. The most amazing sound, apart from her children, was hearing bubbles in a gin and tonic.

Previously she had one of the best digital implants in the world, but could only hear 10% of what she can hear now. The implant has been life-changing in that it has given her a much better picture of the world. It has not been easy though. Fenella's brain has never had to learn to screen out background noises, and I'm having to learn to screen out unimportant ones and pick up the important ones . . . but the learning is slow.

During the interview, a dog barked in the background and she immediately turned around to look over her shoulder – proof if it was needed of the effectiveness of the implant.

Gin and tonic has seldom given more pleasure. While most enjoy drinking it, Fenella listens to its singing first!

©ISTOCKPHOTO.COM/SPANISHALEX

'Sorry – it's a bit noisy', she said, not spotting the irony.

The interesting thing here is to see how much someone's life changes when you give them a sensory experience that they had never owned before. I'm not sure whether Fenella is an enhanced human, because the implant has just moved her towards having normal hearing, but it goes some way to revealing that if we ever do plug in a new sensation, life could be very different.

A Tragic Afternoon

Life changed again very suddenly one afternoon. While on a holiday in Italy in 2006, Fenella and her family were staying with two other families in the hills near Siena. It was their second day, and there was a heat wave affecting the area. Having spent the morning at an outdoor pool they decided to go and prepare some lunch. The families had just taken a group photo and were tidying up clothes when a couple of large raindrops fell, followed almost instantly by a lightning strike that knocked most of them down.

Fenella came to but couldn't hear. Her implant had stopped working, but after a minute it rebooted and she could hear a child's voice whining and calling for her. The child was shocked, but all right. Her husband, however, was dead. Fenella returned home the next day to face the press and a funeral. At this point the implant's value redoubled. She had to deal with the funeral, handle phone calls and now cope with all of life's complications such as talking with the children's teachers. People who are deaf cope, but for Fenella, the implant came just in time. From her point of view, every deaf child should have one, as the new sensation helps her make more sense of the world.

REPRINTED WITH PERMISSION FROM FENELLA

'This was taken in Sydney – the implant enabled me to have the courage to take my kids round the world for a month in April this year – I did take a lady friend to start with but then realised that I could manage on my own and that it helped me to pick up the English language spoken by different nationalities! Almost to the point whereby I had to help other English guests when they couldn't understand!'

And Then, for Something New

Undoubtedly, cochlear implants help people hear. The technology forms a bridge between the sound outside and the sensation in the brain. But why should the technology be limited to picking up sound? Most normal hearing aids have a 'T' setting. Flicking a switch on the back of the receiver turns off the microphone that picks up vibrations in the air, and turns on a radio receiver. If you are in a building equipped with a hearing loop, this means that you can hear the information broadcast through the loop, without getting any of the background noise. It is a great help to people in public buildings like theatres, churches and banks where the shuffling of feet and murmuring of other people can make it hard to concentrate on the person you came to hear.

So why not do the same with a cochlear implant, except why limit it to picking up signals from a hearing loop? Why not build a mobile phone directly into the implant? All you would need is a touchpad to type in the numbers, and with voice recognition software that could be done away with. That said, you may want to keep the transmitter away from your brain, if you don't want it sautéed.

Kevin Warwick suggested moving one stage further. Think of a world where we have worked out how to analyse brain waves so that we can determine what people are thinking, or have trained people to create distinct brain wave patterns. Now, how about recording one person's brainwave information, and sending it directly to another person's cochlear implant? You could have a whole new form of non-verbal communication. Send the signal via the internet and the two people could be thousands of miles apart, and still aware of each other's thoughts. It's an interesting possibility, and would provide a whole new mode of brain-to-brain awareness. The technology at the cochlear implant end of the communication seems to be ready and waiting – time alone will tell whether the brain wave detection systems will ever get good enough to supply the required information.

Two people could be thousands of miles apart, and **still aware** of each **other's** thoughts

IV Faster than Life

Competition is fun. We use it for motivation in many areas of life, not least in sport and in business. It nurtures qualities such as perseverance, stamina, strength and co-operation. It often enables individuals to push themselves to the limit and work as teams. It shows humanity at its best – but the desire to win can also show it at its worst. The drive to be top is all too easily turned into a desire to vanquish whatever the cost.

Human enhancement technologies impact human performance on four levels. There are technologies that are close cousins to therapy. They give new abilities to someone who has lost his or her ability to perform some task, whether before birth or through a disabling disease or accident. Other technologies improve the performance of people who have average abilities to give them exceptional abilities. Then there are technologies that stretch the human body beyond any previously known performance and give it record-breaking powers. And finally there are technologies that enable individuals to do something new that no human has done before. Whether all of these are human enhancements will entertain the philosophers for years, but all have the potential to enhance the lives of the individuals who use them.

In this last section, I want to look for and meet people who have used technology to address difficulties they have faced and to give them a renewed ability to compete in life and in sport. I also want to look beneath the 'drug cheats' and 'gene doping' headlines and investigate the science of some of these drives to improve performance with technology.

When it comes to sport there are a few burning questions. Why do people compete? And what makes sport so important to societies? What is a fair competition? Will our ideas of this change science and society?

10 Restoring Function

Born on 25 September 1952 in New York City, Christopher Reeve achieved world-wide fame as the all-conquering superhero Superman, a comic book and silver screen human-like being with superhuman power. But the actor's fall from a horse during a cross-country competition on 27 May 1995 cruelly showed that when the special effects were no longer playing, he was very human. The accident broke the top two vertebrae in his neck and it was remarkable that he wasn't killed – instead, he was paralysed from the neck down, even requiring a ventilator to keep him breathing. His persistent hope, right up to his death in 2004, was that science would come to his rescue, repair the crushed nerves in his spinal cord and set him on his feet again.

If you have the misfortune of damaging your back, the lower the damage occurs the more function you get left with. A mid- or low-back fracture will tend to leave people able to breathe and use their arms, but in need of a wheelchair to get around. This reduced level of disability gives more scope for developing technologies that may get people walking again. Quite clearly the ability to walk would seem to be an enhancement for anyone who was previously confined to a chair.

One of the most experienced scientists in this area of research is Professor Nick Donaldson. He works at University College London, and I called his office to discuss what hopes there are now for people with spinal injury. I'd written about his work back in 1997, three years after he had been involved in implanting electrodes into the nerve roots of a patient called Julie Hill, who had broken her back in a car crash. I was interested to see how the work had developed since then.

It had all the hallmarks of a straightforward engineering challenge

When he started his research his aim was to restore walking for paraplegics, to take people who had had accidents or illnesses that prevented their legs from working and fix the problem. It had all the hallmarks of a straightforward engineering challenge. Through the 1970s and early '80s he set out to build implants that would survive for years once they had been placed in the body. He hoped that by the time he had developed the implant, someone would be able to tell him how to use it to make people walk.

'I suppose that was my view for about the first five years, when I was extremely young and naïve', he laughed, but time provided a reality check that ended up redirecting his research.

Our legs work via a complex communication network. A message from the brain travels down the spinal cord. It leaves the side of the cord at what are called nerve roots and travels on through peripheral nerves to the muscles. Information in stretch receptors in the muscle is then fed back to the spine. Here nerve cells work out how much more to make the muscle contract to achieve that task required.

In his early work in the 1970s, Donaldson implanted stimulators that had just six channels and gave a very crude function. The people could stand up and swing their legs through using crutches, but it was a long way from walking.

At that time other researchers were trying to solve the problem by sticking fine wires through the skin, using them to channel electrical pulses straight to the muscles. One team in Cleveland, Ohio, put in up to 80 at a time, and then connected the best 30 to a computer. This went a long way towards getting people walking, and although it was an automaton-like walk, Donaldson was quite impressed. However, the Ohio team's patients weren't as impressed, and after a few years some started to sue the researcher because of the infections they got from having so many wires stuck in their legs.

In the early 1980s, Donaldson concluded that the solution lay in an implantable system that could communicate with a large number of muscles. He put an implant into one patient which had 28 electrodes and six metres of cable. Some linked to nerves, some to muscles. It was a pretty heroic piece of surgery, and worked quite nicely for a few weeks.

A key problem with anything you implant in the body is the risk of infection. If bacteria start growing in the thin layer of liquid that hugs the side of the device, then they will multiply rapidly. Within a couple of days you can have a serious problem. On top of that, the bacteria form rafts and colonies that are then capable of resisting antibiotics. Sadly, the volunteer got an infection, and all of the wires and electrodes had to come out again. It took a day and a half in the operating theatre to get it all out. 'It was such a colossal failure', said Donaldson, 'that I just felt this was not going to be the way forward'.

A key problem with **anything** you implant in the body is the **risk of** infection

The next idea was to target the nerves where they leave the spine. By putting electrodes on these nerve roots, Donaldson hoped to stimulate many muscles without having to tunnel cables throughout the body. Operating on the spine is never simple, but relatively speaking it was less invasive than their previous attempt. The patient who volunteered to try this was Julie Hill. She had broken her back in 1990 and by 1994 was ready for a new challenge. The surgical team exposed the nerves as they exited her spinal cord and fitted electrodes to 12 nerve roots.

Footprints in the Snow

Julie's story was written up in the book *Footprints in the Snow* which in 2005 ITV turned into a TV drama. In the film, Julie was played by Caroline Quentin and Nick Donaldson by Jonathan Aris.

'It's never fun seeing yourself turned into a caricature for a film', commented Donaldson. 'TV script-writers seem determined to make engineers look like idiots!'

Then came a long, slow process of learning how to stimulate the nerves to get something approaching the desired effect. Julie and the team worked extremely hard and after months of training she could stand up – in a fashion. It was difficult because she could either extend her hips or her knees, but couldn't do both at the same time. She could stand but only with her bottom sticking out awkwardly and this meant that the implant had limited genuine function.

The breakthrough for this whole area of work came when Nick had what he describes as a revelation – a moment of inspiration. It dawned on him that

❝ My view now is that you should forget about walking and get people on a trike ❞

restoring natural movement was a step too far. It was not going to happen. But enabling a person to move could be possible – if you built a pedal-powered trike. 'My view now is that you should forget about walking and get people on a trike', says Donaldson.

Julie's trike was designed so that she can sit on a reclining seat, with her feet on some raised pedals. The pedals have a sensor that sends information about what point in their rotation they are at any moment to a small computer. The computer then sends signals via her spinal implant to the relevant muscles, synchronising

After Julie broke her back, technology enabled her to break new barriers. Mixing careful surgery, state-of-the-art computing and careful engineering gave novel possibilities.

the way they contract and relax. It wasn't long before Julie could head out on flat roads and pavements under her own steam.

> ❜❜ I felt that **the bike** and the computer became part of me. It was my **muscles** and **my legs**, and then the computer was an **important part** of my making the **bike** travel down the road ❜❜

This has huge benefits. First there is the achievement of moving yourself around, which was after all the initial aim. Then there are the health benefits of exercise. People in wheelchairs get limited access to exercise and as a result become unfit and then are prone to suffer all the heart and circulatory problems that any unfit person faces. Electrical stimulation is the only way to exercise paralysed muscles. Exercise is important because flaccid muscles are unhealthy and cause secondary medical complications. Keeping muscle bulk in the buttocks is particularly important as this stops the bones pushing through to the skin as you sit in the chair all day and causing pressure sores. Not only are these uncomfortable, but they can become infected and threaten life. Now there is the possibility of giving people something useful without the need for trying anything as complex as standing or balancing.

'I felt that the bike and the computer became part of me. It was my muscles and my legs, and then the computer was an important part of my making the bike travel down the road. It was a wonderful thing', said Julie.

Waiting for the New Year

Julie had her accident in 1990, the implant in 1994 and learned to pedal her trike by 1997. I decided to find out how things were going in 2007. I phoned her. I'm not sure whether I was surprised or disappointed to discover that she hadn't made use of the implant for the last four years – probably both. It wasn't that it had stopped working, or that she regretted having it in. It wasn't that she found it cumbersome to use, or that it was massively time-consuming to set up. Julie simply decided that she needed a break. 'It was a matter of taking charge of my disability', she said.

Julie seemed to be facing two issues. One was that she felt she needed to stop fighting the disability and learn to live with it, and the other was the same sense of inertia that hits so many of us when it comes to looking after our bodies. For a paralysed person to be able to exercise is a great health benefit, but like any other exercise regime, it is hard work to keep it up consistently. Right now I plan to go swimming every morning. I've been planning to for the last few months. I know it will do me good, I know I'll feel better for it, and each day I decide it had better wait for tomorrow – after all I've got a book to write. There just seems too much to do. It's Julie's equivalent of 'I'll go to the gym tomorrow'.

But that wasn't the full story. Julie also had a note of disappointment. Yes, she was installed with the most advanced functional electrical stimulation (FES) system on the planet, and yes by the standards of FES it was working well. But she was sad that the device hadn't progressed further. She felt that she had worked as a prototype to help others, but hasn't seen it progress.

'I don't regret any of it, and have the deepest respect for what the team did on a shoe-string, I just wish the research was going further', she told me.

She was also affected by the hype that accompanied her first trips on the bike – the news coverage, the book, the film. But for her it was more of an exercise machine than a tool for daily living. It was a prototype, and so was somewhat Heath Robinson, and there had been three years of hard work between the implant going in and being able to cycle.

All the same, Julie hasn't given up hope of getting on her trike again. She still has the controller and could in theory just plug it in and press the on button. I say in theory, because right now it is packed away in storage as she and her husband are in the last stages of emigrating to Australia for a few years. She hopes that once they arrive in Australia, she will have a New Year's resolution moment and get going again.

I'd written one of the news stories that Julie mentioned. It, too, had been upbeat about the new world that was just around the corner, and hopeful about the new possibilities that this technology raised. For example, Julie did her initial muscle training lying in bed, with the stimulator triggering the muscles. It was tedious and physically tiring, but she could doze off as it went on – just think of that, exercising in your sleep. And like many I'd asked not only whether this was the moment when disabled people's lives were about to be enhanced, but whether the same could be true for the rest of humanity.

Surface Electrodes

For the past five years, Donaldson has been focusing on squeezing as much as he can from surface-mounted electrodes, and getting people to either use trikes or work at specially equipped rowing machines. The system is easier to fit to a person than surgical implants, and is still good at toning up muscle, although the power that the muscles can develop is not huge.

They can produce spurts of perhaps 25 watts, but then this quickly falls to about 15 watts which they can produce continuously for hours. But that is considerably less than the 80 or 90 watts continuous power that an able-bodied person can readily put out, increasing to around 150 watts when you meet a hill.

The **challenge** is to **provide** more **benefit** than hassle

The 64 thousand dollar question is 'What is limiting the power?' Donaldson says that stimulating with surface electrodes simply doesn't seem to recruit all the motor units in the muscle group that you want. You might get 70%, but the other 30% do not respond. If you could stimulate the nerve, you'd have a better chance of stimulating all the motor units, but that is much more difficult to achieve. The challenge is to provide more benefit than hassle. It currently takes around an hour to place the 16 electrodes on before someone can start to use an FES system, which means that the benefit needs to be great if a person is going to strap it on four or five times a week.

Benefit can come in many different measures. Donaldson says that some people are happy cycling slowly along outdoors, particularly if it's a sunny day. Others find that boring. For them, indoor sport could be a pleasurable alternative. Here FES rowing is a distinct possibility. Just as you can equip a trike with sensors telling muscles when to contract and relax, you can do the same with a rowing machine. The number of FES oarsmen and -women has risen to the point that they are beginning to organise international competitions.

Functional electrical stimulation is setting free many back-damaged people to engage in exercise and sport. It enhances their lives, but doesn't yet push performance beyond 'normal'.

Christine Spray

Christine loved the outdoors. She used to help her partner run a commercial shoot, and also rode horses. It was falling from a horse in 2000 at the age of 40 that broke her back.

'I read an article talking about FES experiments and immedately wrote off to see if I could join in', Christine told me. After a bone density scan showed that her bones were still strong enough, she joined in with the programme.

It took just a couple of hours to fit her out initially with the 16 surface electrodes that stimulated her leg muscles, and she could start cycling straight away. For 10 weeks she did leg exercises lying in bed and then built up to five hours a week. At the moment she uses her trike on an exercise frame in the house because the country roads outside her house are too rough for the trike.

'It's a tiny bit bizarre to see your legs moving – you think "they can move", why can't I do it. It's very difficult to describe . . .'

'And you keep that going now?' I asked.

'I hate to admit . . . well not quite', she replied. 'We all think, I must exercise, and spinal injured people are just like able-bodied people and put it off. After the initial year it has dropped to about half of that.'

Christine has recently been involved in another research project that involved doing 10 minutes' exercise with weights attached to her legs. This has strengthend the muscles, but has also increased the number of times they go into spasm.

'Hopefully next spring I might get out on the trike – ten miles would be a great achievement to aim for.

'I've never thought about the machine being part of me – I'm just so pleased that there are able-bodied people out there trying to help less able-bodied people.'

Robotic Walkers

The only alternative to FES sports approach would be to turn people into robots – to put loads of miniature transducers in every muscle and wire them all up to a complex controller. Although Donaldson thinks that this is theoretically possible, he believes it would be a colossal waste of effort. His experience working with people who have different forms of paralysis has shown him that most people want to rehabilitate, to be as much as possible like normal people, not turned into robots. His aim now is to make a Julie-type implant which, along with the exercise, controls bowel emptying and bladder function, and can also produce erections in men if they want. Then, he thinks, you would have something that was worth using as part of normal life, and not just as an advanced exercise machine.

The **result** would certainly enhance damaged bodies, and I think, enhance the **lives** of those with **damaged bodies**

The technology is basically there, and an implant could be a matter of three or so years away. Unless there was any objection from the regulators, it's a matter of raising the money and of finding the surgeons to do it. The result would certainly enhance damaged bodies and, I think, enhance the lives of those with damaged bodies.

Brain Interface

I wondered if Donaldson had ever thought of bypassing a damaged spine by using a brain implant to collect the signals.

'I just occasionally hear someone talking about brain interfaces at meetings and I think generally what strikes me is how little progress they've made', he replied. 'I remember talking about this when I started, 30 years ago, and you know, all the same sort of issues were clearly there. There was no obvious technical reason why we couldn't place electrodes and arrays on the brain then and there didn't seem any reason why one shouldn't be able to get bit rates up to tens of bits a second rather than the one or two bits per second achievable now. Really that is no better than pointing at a board with a laser thing or getting a camera to track eye movement so that you can type at an on-screen virtual keyboard by staring at letters one at a time.'

Reeve's FES

The media headlines that accompany each new development in FES talk about the era of the bionic man, raising expectations of restored or even improved function. But as I look at the best that is available, it is clear that that is not even on the horizon. The irony of Superman, actor Christopher Reeve, falling from his horse and becoming almost totally paralysed should stand as a reminder of the scale of the problem. But at the same time, his story also has a glimmer of hope.

Christopher died of a chest infection that couldn't be controlled, but before that he had shown measurable improvement. He was regaining control of some of his muscles. The interesting thing is that while Christopher was a vocal proponent of stem cell research and therapy, it was FES that made the difference for him.

Five years after his injury, Reeve started to use an FES bicycle and built up to one hour per day, three times a week. Once his muscles had built up strength he started exercising in a swimming pool for one hour each week. This aqua therapy focused on using the muscle groups in which he was beginning to get some restored voluntary control. Working in water is great, as even small movements can be detected because the water supports your weight. Once he detected movements he could work on them further.

After three years, Reeve was re-building the ability to move muscles and sense when things touched him. It was a long way from a cure, but it was showing that if you stimulate the body, it has a much better chance of recovering. Scientists monitoring his progress speculated that this could have been because a few nerves had survived the injury and could be reactivated. Another option is that new nerve connections grew across the break.

After three years, Reeve was re-building the ability to move muscles and sense when things touched him

'It's strange', says Donaldson, 'He talked to the public a lot about stem cells, but seldom talked, as far as I know, about the fact that he was doing FES cycling with his medical team, and he probably had more FES training than almost any other patient in the history of spinal cord injury.'

Not only had his **muscles grown stronger**, but he had **learned** to control the **muscles** in a way he couldn't **before**

A patient that Donaldson is used to work with had a similar experience. He was partially paralysed but before his accident had been a cyclist before his injury and wanted to get back to cycling. He was also an ex-GP and understood that it was good to exercise. Donaldson's team gave him a stimulator and he went home and used it regularly, often while watching a film. After 15 months they were amazed to find that he had relearned to use his more paralysed knee and could now produce strong knee extensions. Not only had his muscles grown stronger, but he had learned to control the muscles in a way he couldn't before.

One exciting possibility for therapeutic use of FES is that this could be due to re-mapping in the brain or re-mapping in the spinal cord. It could be that because the stimulation is phased at the same time as the legs move there's a sort of a relearning process. The subject said it made significant improvement to his ability to get around. He often walked with two walking sticks and he could do things like pick objects off the floor, which he couldn't do before.

Bionic?

The press is very tempted to write up stories like Julie's and Christine's with a title that includes the word 'bionic'. The impression is that we have the makings of something superhuman. As far as I can see, the reality is a technology that is on the verge of delivering an exciting therapy – but let's hold back from any claims of enhancement.

11 Doping – Drugs and Genes

Chariots of Fire **is based on the** true story of two young men as they prepared for, and competed in the 1924 Summer Olympics in Paris. It carefully depicts the struggles and pressures placed on athletes by themselves, their families, friends and at times their nation. It also charts an interesting development in the history of sport – the notion of taking training seriously. We see two teams pitching themselves against each other, one with rigorous schedules, the other with filled champagne glasses balanced on hurdles. One of the many stories that the film tells is about professionalism's entry into the Olympics.

It was no longer good enough to be born fast; you needed to take that in-born ability and stretch it. Sport now faces a permanent and growing issue – what is stretching a natural ability and what is cheating by introducing some unnatural element?

No large sporting event goes by without the accusation of 'drug cheat' being thrown at someone. The 2007 Tour de France was a good example. By the end of the Tour, two cyclists, Alexandre Vinokourov and Cristian Moreni, had been dismissed for testing positive for a banned substance – Vinokourov had been the pre-race favourite. The next day, the wearer of the coveted yellow jersey, Michael Rasmussen, was taken out by his team after their sport director said that he had been less than truthful about his whereabouts when he missed doping tests. Later investigations found that Andrej Kashechkin and Iban Mayo had tested positive for banned substances during the last stages of the race. On top of this, two teams, the Astana Team and Team Cofidis, were asked to withdraw due to at least one member's positive test results.

Alexandre Vinokourov pushes hard to win. But tests then showed he had performance-enhancing drugs in his body. In the eyes of the sport's governing body he went from victor to public villain in one quick step.

We call them cheats because they have broken rules. The rules are set to enable competition on level terms within agreed levels of safety and risk. Rules relating to drugs attempt to prevent anyone from taking a chemical that is not normally found in the body or obtained from standard food and using it to give a level of performance that they could not achieve by training alone – a level that for them would be more than human. If, or rather when, people do this, their personal ability is certainly enhanced.

The rules came into being not out of a desire to spoil people's fun and limit achievement, but to keep people safe. In a liberal society we try to use restricting rules and laws to curb people's freedoms only if this prevents them from hurting themselves or others. The issue is that if drug taking becomes standard, then everyone will need to do it if they have a hope of winning. But drug taking is not safe. Indeed a pivotal point in the history of athletics was the Olympic Games in Rome 1960 when Danish cyclist Knud Enemark Jensen died during competition, and a post-mortem examination revealed traces of amphetamine. Taking performance-enhancing drugs had pushed his body beyond its safety limits and caused catastrophic failure. The short-term gain had been very short. At first sight this may look as if the law is protecting would-be drug takers. In fact it is protecting those who don't want to be forced to join in with the drug-users in order to have a chance of winning.

One problem for sports-governing bodies in general, and the World Anti-Doping Agency (WADA) in particular, is trying to keep one step ahead of the 'cheats'. Science and technology combine to produce a gentle stream of new ideas and new products, and the sports organisations are then left trying to develop tests to see if they can catch people using them. Whether you have enhanced humanity is more questionable. Taking a drug will change an individual, but it won't forever change the species.

Introducing EPO

What happens, however, if athletes enhance their capability by boosting something that their bodies normally have on board, not to extreme levels, but just to the level that would be seen in a few lucky but normal individuals? It would be almost undetectable, and would give them a competitive edge.

A key example of this is EPO (see box). This hormone is manufactured as a therapy, but also taken illegally by athletes. Some people claim that EPO injections are commonplace in international sport. The problem in assessing this claim is that it is based on anecdotal evidence. To use it, athletes need a constant supply of the substance, as well as syringes and needles that would need to be kept well and truly out of sight of any inspectors.

EPO

In order to mature, baby red blood cells in bone marrow need a molecule named erythropoietin, or EPO for short. Red blood cells are important because, among many things, they carry oxygen around the body. The more EPO you have in your bloodstream, the more red blood cells are likely to be released from the bone marrow, and the more oxygen your blood can carry. More oxygen equals greater stamina.

EPO is normally produced in the kidney, and various people have been looking for artificial means of introducing it into people who have different forms of kidney disease.

Artificially it is made by placing the gene inside cells growing in flasks full of nutrients. The cells kick out EPO, which can then be harvested and given to needy people – or taken illegally by sports folk.

Currently, WADA has a campaign against EPO and puts a considerable effort into developing and performing tests that aim to distinguish among the different methods an athlete might have used to increase oxygen levels. Their aim is to exclude not those who are in danger from overly high levels, but those who are not relying on their kidneys to produce the stuff in the normal way.

An ethics expert at the University of Oxford, Professor Julian Savulescu, has pointed out that this strategy has a flaw. Some families have unusual gene mutations causing their members to carry a variant of a gene that builds the EPO receptor on the surface of the cells that generate red blood cells. The consequence is that these people produce higher quantities of red blood cells than most other people.

The superbly successful Finnish cross-country skier, Eero Mäntyranta, was a member of one such family and the key subject of an important study into the area. This genetic explanation was only discovered in the 1990s. Eero competed in the 1960s. Would he still be allowed to compete today?

Eero Mäntyranta broke records because he worked hard, but he was also aided by an unusual genetic makeup. Was this an abnormality or an enhancement?

Some key moments in Eero Mäntyranta's record of skiing cross-country
success

Olympic medals   Three gold
                                         Two silver
                                         Two bronze

World championship   Two gold
                                         Two silver
                                         One bronze

National domestic championships   Five gold
                                         Two silver
                                         Two bronze

Having an unusual EPO gene is not a magic bullet for sport. Mäntyranta still felt enough pressure to perform to take further risks with his body. He was suspected of using amphetamines in the late sixties, although he denied it. He did admit to using hormones, which were at that time legal. Although he lives to tell us about it, he could quite easily have died early, as some top cyclists are reputed to have done because of an increased red blood cell count.

The availability of EPO and confusion over whether it has been created by the body or injected is one reason why Savulescu claims that now is the time to abandon any inhibition on drugs and let athletes have free reign to take what they want as long as crucial indicators of health are very tightly monitored. For him, it is dangerous to an athlete's health to compete with elevated EPO and too many red blood cells, and it does not matter whether those levels are natural or not. The athlete's safety should be the prime concern of sports' controllers, and competitors showing up with higher than safe levels should be prevented from putting themselves at further risk by competing. This escalation of regulations would, however, go against the libertarian principles that drive most of enhancement's desire to 'live and let live'.

It is also simplistic to think that we can always spot which health indicators are important in the long term. In 1976, the East German swimming team put in a magnificent performance at the Olympics, winning 11 out of 13 events. The athletes looked like superb examples of healthy human beings. But their success came at a cost, and they later sued their government seeking compensation for the damaging effects of the drugs they were forced to take.

So even Savulescu's apparently open-minded view runs into trouble as soon as you move it out of the academic's starting blocks and let it run for real – many, if not most, drugs will still be banned because of their threat to health. Thinking about it, how many sports will also get banned? Rugby breaks arms, legs and backs; football takes out the metatarsal with remarkable frequency; and as for downhill skiing . . .

On the other hand, letting people tune their EPO levels would lead to more people competing at the top limits of human performance, but would not lead to a situation where we have any increase in the upper limit. Given that taking many of the potentially enhancing compounds breaks no civil laws, there is no reason why we won't see a set of 'extreme sport' events in which everyone knows that the competitors are risking all in their desire to win. Extreme sports have their competitors and spectators, but they tend to be a select crowd. Most people think there is something distasteful in watching a pursuit that has a high chance of causing severe injury to the participants. To my mind that would be a bit like returning to the Coliseum where we could watch people perform in a way that tore their bodies apart. The difference would be that we would probably be spared the gore, but the effect on the performers would be similar. And however EPO is used, I can't see it generating a truly enhanced human.

My Myostatin

Before we concentrate too much on taking EPO, we need to realise that this is only one of a range of options, and many involve one or more of hormones that regulate energy flows in the body and growth. A team of researchers at Johns Hopkins University in the USA are studying a molecule that they have called IGF-8. IGF stands for insulin-like growth factors, and IGF-8, or myostatin as it is commonly called, is the stuff that keeps satellite cells inert in adults. Satellite cells nestle within muscles ready to leap into action and transform into a muscle cell if physical activity demands an increase in muscle mass. They are basically stem cells. Myostatin acts as a brake.

The more myostatin you have floating around in your blood, the fewer satellite cells transform, and the less your muscles grow and rejuvenate. Now it turns out that farmers have been breeding cattle with reduced myostatin levels for generations. In the Piedmontese breed, from a remote area of northern Italy, farmers began deliberately to select very muscular animals. From the 1880s they began to breed powerfully built cattle that have become known as 'double-muscled'. The meat was particularly lean and there was more of it than usual on a slight bone frame.

In a similar way, Belgian Blue cattle have come about as a result of intensive breeding of exported British shorthorn cattle that went to Belgium in the early twentieth century. Muscular traits were encouraged during the 1950s and in the 1960s true double muscling appeared.

Of course it is only recently that the genetic basis of this has been found. In Piedmontese cattle, for example, there is a single letter transition (from G to A) in the cattle-equivalent of the myostatin gene. As a result these doubly muscled cattle fail to produce myostatin at all.

Banned Beefy Whippets

The system is not confined to cattle. A team of US and UK researchers studied the myostatin gene in racing whippets. Of the 41 dogs they tested in the top performing classes, 12 had a gene mutation that left them with less myostatin. In the lower classes, only one in 43 dogs had the mutation. For a whippet to win at racing, it needs to have rather less myostatin than usual. Intriguingly, whippets with no myostatin at all have a problem. They are so over-muscled that the dogs no longer meet the Kennel Club criteria for the breed. Consequently they are normally removed from competition, and the researchers only found one such dog in regular competition.

Curiously, in racing greyhounds, which are genetically very close to whippets, the researchers did not find an equivalent mutation. They speculated that this was because greyhounds race over a longer track and thus need more endurance. In mice myostatin-deficient animals have more 'fast twitch' type II and glycolytic fibres in their muscles, which are the ones that help you sprint. They had fewer 'slow twitch' or type I and oxidative fibres which give you endurance.

Should We Be Worried About Becoming Myostatin-ist?

Would lower myostatin help or harm humans – enhance or encumber? Well, there are over a hundred recorded cases of adult humans with myostatin deficiencies. They are taking part in a study at Johns Hopkins University hospital run by Dr Kathryn R. Wagner. This was instigated after two cases of infants with myostatin deficiency came to light. The first was a German child who has a change in the gene coding for myostatin. He was producing no myostatin at all and the boy's family tree showed signs that other relatives were partially affected by the mutation. His uncle and cousins were known for their exceptional strength, as was his mother's father and his father before him. His mother had been a professional athlete.

A second case was Liam Hoekstra. Liam's mutation is such that even though he produces myostatin, his muscles can't recognise it and so they keep growing. Although still a small child, his muscles are dense and fast. He eats a full meal almost every hour of the day and can already move heavy furniture.

He loves to run, and has almost no body fat. In fact doctors worry that his brain and nervous system will not form correctly because they need at least some fat to develop.

If you want to make use of this knowledge to boost your performance, then how about taking a drug that inhibits myostatin's actions? They are in clinical trials and are likely to enter the marketplace. Indeed, the pharmaceutical company Wyeth, in collaboration with Cambridge Antibody, is already testing a myostatin-based treatment in human trials. Used in otherwise healthy people, myostatin inhibitors might produce an increase in muscle bulk in quite a short time – they certainly do when injected into mice.

Alternatively, if you and your partner are both sporting successes, there is a chance that you could have genes that give you low myostatin levels. How about using fertility techniques to screen any embryos you generate to search for those with genes that would stretch this deficiency further? If this worked, would you tell the child, or should the child be only allowed to compete in categories of similarly selected individuals?

What if a fertility company started buying sperm and eggs from Olympic champs and selling screened embryos to the highest bidders? When the child grows up, should he or she be allowed to compete, or would this person simply be born to cheat?

Who's Watching?

Sport is often thought to be about fair play, trying hard and taking risks. Some commentators have said that athletes using new technologies would be trying harder and taking more risks than those who naturally had these gene sets. Surely they should be given more respect for their efforts?

It's a sentiment that has been argued over the summer of 2007 in the journal *Nature*. An editorial published in August argued that bans on drug enhancement in sport were a waste of time and should be removed. The basic argument was that banning drugs came from the same mindset as did former generations who prevented women from competing, or who stopped athletes from winning money. It asked, 'Is it really reasonable that athletes should make do with bodies that have not been enhanced?'

The reply came in the letters pages a few weeks later. One correspondent said that the editorial missed the point – 'you fail to recognise the reason why people drive 10 hours to watch the regional final of college basketball, wake up in the middle of the night to watch the inevitable penalty shoot-out at the end of an England World Cup football match or even hop on the fast train to see the yellow jersey lead the Tour de France . . .' People, he argued, love sport because they can marvel at how great talent can take human achievement to its limits – there would be little appeal to watching drug-enhanced people. Personally I think there would be some appeal, just as there are fans of body-building competitions in which entrants have clearly taken drugs to create extreme physiques – but I don't see it taking over the mainstream arena.

Repoxygen Raises Eyebrows

Up until now we have been looking at ways of manipulating the body with various forms of external agent – a drug or hormone. But how about manipulating the body so that it produces its own drugs or has altered hormones? How about introducing into the body novel genes that have a similar effect to a performance-enhancing drug? This would be particularly the case if someone decided to place the genes in an early embryo so that they became incorporated throughout the body and passed to any children he or she may have in the future. Here we could see the arrival of an enhanced human, maybe even enhanced humanity.

Some people **are claiming** that modifying **genes** might just **be starting to creep** into sport for **real**

It's been the stuff of fiction and fanciful headlines for decades, but some people are claiming that modifying genes might just be starting to creep into sport for real.

For athletes who want to use EPO there are a few inconveniences to overcome. You need a constant supply of EPO and you need to check your own blood levels. Both are issues that require working with others. There is always the risk that someone will 'tell'. Then you need your needles and syringes and somewhere to hide the lot.

You could, however, modify your body to produce additional supplies of EPO. Repoxygen™ is a gene therapy that aims to do just this. Developed by a UK-based drug development company, Oxford Biomedica, to treat anaemia, it is the commercial result of research carried out over a number of years at Pennsylvania University. This therapy was created after scientists tracked down the gene that causes kidney cells to build EPO and linked it to a gene that will turn on EPO production if cell oxygen levels begin to fall. They placed this genetic cassette inside a virus and used the virus to transport the gene into muscle cells. This would modify local muscle cells in that individual, but not affect eggs or sperm, and consequently not be passed on to any progeny.

To test it, the inventors injected this novel gene-containing virus into the leg muscles of mice and watched the EPO levels rise when oxygen levels dropped.

The system worked. The company that manufactures it makes no claim about any trials in humans, but if you read the news cuttings about it you would think it was on the pharmacy shelf already. Despite a lack of evidence that anyone has actually used Repoxygen or any similar therapy, this is being proclaimed by some as the moment that gene doping arrived in sport. As a consequence, a few people are already calling for us to change all our rules, and possibly our notion of fair play.

They all wanted to be bigger, better and more powerful

Part of this extreme reaction has come from research that has received high levels of publicity. When H. Lee Sweeney first showed that his version of gene therapy could enhance mouse muscle, the global news media ran the story as if genetically modified athletes were here. Within hours Sweeney was flooded with calls and emails. Some people want to reverse muscle degeneration caused by disease or ageing, but around half of the pleas came from healthy individuals – professional power lifters, sprinters and weekend wannabe athletes. They all wanted to be bigger, better and more powerful. It's worth noting that the work has not successfully moved beyond mice.

Another incident added fuel to the fire. In 2004 the German athletics coach Thomas Springsten was sacked from his club. He was accused of using banned substances with the athletes he was training. As police searched for evidence they came across an email in which Springsten was lamenting the difficulty he was experiencing in getting hold of Repoxygen. There is no indication whether he had already obtained and used it, or was looking for it ready for a first use, but just the thought of someone seriously considering using this product sent shockwaves through sport.

Despite any evidence that Repoxygen is available, the claims of imminent gene doping have been made with such vigour that WADA has had to react. In their desire not to be overtaken by events, they have set up a special funding stream for research that aims to build tests that could detect whether someone has added novel genes to his or her body.

Potentially Enhancing

To see whether or not WADA need to be worried, it is worth having a closer look at Repoxygen, as it is held up as one of the leading contenders. Repoxygen works by genetically modifying muscle so that it produces EPO if oxygen levels in the tissue fall. This gives a chance of enhancing the lives of some people whose kidneys aren't producing EPO properly, as it removes the need for frequent unpleasant injections.

For an athlete, it is difficult to see the attraction. Treating someone who has a full blood count with Repoxygen would have a negligible effect. It has a built-in sensor that only activates the genes if oxygen supplies fall dangerously low. If the person is healthy, then training will naturally build high oxygen-carrying capabilities and muscles would only be depleted after extreme exercise. Consequently the new genes would rarely get switched on. Even if it was available and safe for human use, Repoxygen doesn't look like a huge threat to conventional sport.

Sportoxygen

I guess you could engineer a new version of Repoxygen, maybe call it Sportoxygen, with a higher threshold at which the gene switches on. This would possibly give rise to a constantly higher level of red blood cells.

There are two problems here. Giving an artificial supply would probably turn off the natural one, so the new genes would be charged with taking over the system, not supplementing it. As such, there would be the danger that you could land up with too high a level of EPO and too many blood cells. These would impede blood flow in the capillaries, and decrease the amount of oxygen available the body.

Gene Therapy Reality

It is also worth taking a brief glance at the state of genetic therapy in medicine. Here we move from anecdote to a more carefully recorded domain, and the picture isn't too pretty. And at this point I can feel the hypeoscope warming up.

The early hope of gene therapy was that if a person inherited a faulty copy of a gene, then gene therapy would supply him or her with working copies of that gene. This would genuinely cure him or her and if the new gene went into the person's sperm or egg cells, the repair would be passed on to all future generations. In 1992 I went to a press conference in Glasgow. The gene for the protein that is faulty in cystic fibrosis had recently been found and decoded. The protein it made had been identified and characterised. The exact error in the gene responsible for the disease and the effect that it had on the protein were also known. The scientists were justifiably pleased about their work – but it turned out that they were not justified in claiming that within a decade we would have a gene therapy to solve it.

Fifteen years later most people are pretty pessimistic about the chance of gene therapies for genetic diseases breaking into the medical market as cures for genetic disease. There are a number of problems. First you have to find a way of getting the new genetic code into the needy cells. Most approaches achieve this by using modified viruses. The worry and wonder of viruses is that when they infect you they inject their DNA, their genetic code, into your cells. Some of these do it in a way that makes their DNA bind with each cell's central genetic store, the chromosomes. Once fused in the chromosome, the sequence will remain in that cell, and in all future generations of cells that are created when this one grows and splits in two. Other approaches use tiny artificial packets that can carry genetic material into cells – but with these you still need to get your packet to all of the cells you want to rescue. No mean feat.

The idea for cystic fibrosis was to target lung cells, as these were the ones most damaged by the disease. The lungs also provide an extraordinarily large surface area – in the region of half a tennis court in an adult – so giving a reasonable chance of absorbing loads of the novel genetic material if you spray it in. Hitting lung cells also reduced the chance of the new genes making it into sperm and egg cells, and prevented them from being passed on to any progeny just in case the therapy proved unsafe.

Our bodies, however, have good systems for detecting and destroying incoming particles and viruses. In some trial patients, the immune system went into over-drive after sensing the vehicles, and created such an extreme response that the patients died. Even in patients who survived, the body was remarkably good at clearing out these viruses, or destroying the cells that they managed to modify. And it is not always possible to give a second dose. The immune system will often not spot the initial injection of virus until they have had time to do their job. If you do give a second shot, the body is primed and ready to release a rapid reaction force, and once again, that response could go over the top.

After excited hype of the 1990s, almost all major pharmaceutical companies have fled from research aimed at developing gene therapies, and Repoxygen has spent a few years on the shelf waiting like a 50-year-old spinster for a loving partner, and one with plenty of money. Fifteen years after the first gene therapy conference for it, there is no serious contender for a working gene treatment for cystic fibrosis, and with the exception of a few enthusiasts, commentators are looking way into the future before they expect one to break into the market.

In terms of treatments aimed at other disorders, the picture is not much rosier. A recent trial aimed at curing genetic disease involved a dozen children in France who had an immune disorder that doctors tried to treat with gene therapy. Three of them subsequently developed a form of leukaemia as a direct consequence of the treatment. One died.

There is also a little matter of cost. If gene therapy did break out as a success, then it would probably cost in the order of £25,000 per shot. Given that it would probably need repeating every few years, this would be a little outside the scope of most competitors.

Key Steps in Human Gene Therapy

1990	US researcher William French Anderson treats two girls with severe combined immune deficiency (SCID).	Anderson claims success, but for the girls it is highly probable that they survived because they continued to take their conventional therapy.

1993	Boy with SCID treated with genetically modified cells taken from his own umbilical cord.	Treatment worked for four years, and then he needed more therapy.
1999	18-year-old Jesse Gelsinger given gene therapy for potentially fatal genetic disease.	Jesse dies days later as the virus that carried the gene caused an extreme immune response.
2000	French team delivers new genes to 12 children with SCID.	Trial abandoned in 2002 after three children develop cancer and one dies.
2003	Chinese introduce anticancer therapy Gendicine. In 2004 it became the first gene therapy product to receive a government agency licence.	Successfully reduces tumour size.
March 2006	International team treat two patients with a white blood cell disorder.	Results look promising as a potential cancer therapy.
May 2006	US scientists treat two cancer patients with genetically modified immune system cells.	Results look promising as a potential cancer therapy.
May 2007	One patient treated for an inherited eye condition at Moorfields hospital.	Too early to assess outcome.

One of gene therapies' most ardent advocates is Professor Theodore Friedmann, who works at the University of Pennsylvania. In 2005 he warned people about the problems caused by overhyping the current level of success in gene therapy. In the academic journal *MUSE* he wrote, 'Public perceptions of and confidence in [gene therapy] were damaged by the hype. Most unfortunate of all, the hopes of patients and their advocates were disappointed.' Friedmann still passionately believes that gene therapy has a future, but the current data don't live up to many of the heady claims made in the past.

Despite this, it strikes me that he is prone to get overexcited. Against the background of extremely few proven successful clinical trials, his presentation to WADA at a 2005 conference in Sweden that had been called to assess the risk that gene therapy posed to sport was entitled, 'The Irrefutable Success of Gene Transfer for Therapy of Human Disease'. To be fair, there are hundreds of small-scale trials now looking at different agents, but few of these have been irrefutably successful, and the ones that are proving to show most promise are the therapies that kill off tumours. They may make great treatments, but have little application when it comes to enhancement. 'Irrefutable success' is certainly a strong statement for gene therapies' current state of progress.

At first I was surprised that WADA were taking gene doping so seriously at the moment. Then I noticed that Friedmann heads WADA's gene doping panel. He has certainly spent more time than most working in gene therapy, but even there he is known as a controversial member of the scientific community, in part because of the extent of his claims regarding gene therapy. It seems to be that WADA's concern could be strongly linked to his presence in their team.

Mighty Mice

Just as I finished writing this, I picked up the morning paper. The front page was covered in grainy black and white photos of a lab mouse running on a treadmill.

The caption told me that this supermouse could run 20 metres a minute for hours without getting tired. It lives longer, and enjoys an active sex life well into old age. The page also posed the question – can this be applied to humans?

Tinkering with it had produced a mouse with the equivalent capability of Tour de France champion cyclist Lance Armstrong

The colony of research mice had been created by scientists at Case Western Reserve University at Cleveland, Ohio, by altering a gene involved in glucose metabolism. Glucose is the body's main energy molecule, and tinkering with it had produced a mouse with the equivalent capability of Tour de France champion cyclist Lance Armstrong. It begs the question of whether this technique could create an enhanced human, or less controversially create a person who has maximal human abilities.

This treatment didn't come without difficulties. The mice ate twice as much as normal ones even though they were half the weight. Not much of an issue there if you can afford the food, but this energy flow left them highly aggressive. Talking to *The Independent* newspaper, the professor behind the work said 'This is not something that you would want to do to a human. It's completely wrong. We do not think that the mouse model is an appropriate model for human gene therapy. It is currently not possible to introduce into the skeletal muscles of humans and it would not be ethical to even try.'

Despite his concerns, the professor did go on to say that different gene therapy techniques could possibly be used to move the gene into muscles of people suffering from diseases like muscular dystrophy. If that worked and became an internationally used therapy, then it would open the doors to use by athletes. If it did work, it would be impossible to stop 'wrong' use, but the alternative methods of delivery are not available yet, so I don't believe he needs to worry – yet.

Repoxygen's Therapeutic Enhancement

Providing a backup built-in production pathway for an essential hormone might also be useful for healthy folk too. If the natural system of EPO production failed, the Repoxygen would be ready and waiting to cut in. Biological systems often have more than one way of solving important issues, termed redundancy, so that they can survive crises. This gene technology would give a 'designer-built redundancy', a form of engineered reliability. Could this be a new class of enhancement?

And forget sport for a moment. What effects could a product like Repoxygen have on a patient? If you suffer anaemia, you will feel lethargic and depressed. Curing the anaemia would certainly change a person's psychological state, attitude to life and most likely his or her opinion of him- or herself too. As a treatment for one symptom of a chronic condition, this might just help those with anaemia feel less sick and more like themselves.

On the other hand, if they have always been rather anaemic due to having had the kidney problem, they might not recognise their new selves. The issue is commonly discussed in medical ethics, so in itself is not new.

On the positive side, having a built-in supply of EPO might improve confidence because of an increased sense of independence, just as freeing a diabetic from insulin injections improves overall quality of life. But it wouldn't enhance the person beyond what most folk already enjoy.

Them and Us

While I really don't believe that there are any genetically enhanced athletes in current competition, and I am also deeply sceptical that this will change in the next couple of years, you would be a fool to say it could never happen. The question for sport would be how to handle this situation. Having people with genetically enhanced capabilities would exclude from competitiveness those who were not willing to use these or other similar technologies. If you let people use enhancements, you would need to create a new set of classes of competition.

It is conceivable that you could end up with a situation where normal humans race in categories equivalent to historic car rallies while those taking EPO and better race in an unlimited class. To an extent it is nothing new about having categories. We can see the process working in the Paralympics where carefully grading of athletes lets people with similar capabilities compete against each other.

If sport were to go down this route, it would be interesting to see which classes start to assume highest prestige. An assumption from people arguing to legalise enhancement is that the new super race will rule, and that those who don't want enhancement, the 'naturals', should simply roll over and take up as museum relics. This fits with the line that athletics' rules should be modelled on principles that govern mechanically driven sports. The question will be who sets up the new calendar of sporting events. Will the 2012 Olympics be an enhanced event, with a secondary un-enhanced Olympics running as a class within the Paralympic games that run over the following days – or will the Olympics stay within its current frame of expectation, and an enhanced category be spliced to the Paralympics?

Those who **don't want enhancement**, the **'naturals'**, should simply roll over and **take up** as museum **relics**

I'm intrigued that when people talk about letting enhancement into games, the solution is so often to introduce more rules, especially as the transhumanist camp who are most in favour of this tend to argue for autonomy and freedom from restrictions. I'm also intrigued that an assumption made by people in favour of enhancement is that if someone needs to be excluded from a sport, it should be the people who aren't enhanced, the people who got there by struggle and sweat alone.

Enhanced Athletes?

So where does that get us in our search for enhanced humans in sports? I find myself pulled in two directions. The first is an acknowledgement that people are not born equal, and that it is those with extreme traits who tend to win in sport. You could describe the Olympics as the biggest freak show on earth! It is a twenty-first century version of Barnum's circus, a theatre where we gather to gawp at those with strange genetic traits. This isn't to say that they haven't also worked and pushed and strained and trained their way to the top – genes alone will never be enough, but without an appropriate physiology you will be left behind when the starting pistol fires. As such, arguing for a level playing field is complex.

On the other hand, I'm not sure that I've seen many signs of real human enhancement currently in the field. I think we are yet to see any technologies that are altering human bodies so that they can do things that no human being has

Dope-driven athletes are not enhanced – they are simply normal humans breaking current safety rules

ever done before. Gene therapy seems to be on the horizon, but it seems more like the mountainous horizon than a walk in a park. One of the problems when climbing a mountain is that there is a tendency to believe the summit is that furthest point you can see at the moment. But when you get there you find a new challenge between you and the top.

It also seems to me that the modifying treatments are more like therapies that enable the mediocre to compete at the top. Even if we pursue Savulescu's idea of allowing everyone to take therapies that will take them to the edge of human performance, we still wouldn't be creating enhanced humans – we'd just have physically stretched athletes.

It also seems to be a mistake to elevate the status of people who take drugs in order to win by claiming that they are enhanced. The term is loaded with a notion of superiority and benefit. As far as I can see the current generation of dope-driven athletes is not enhanced – they are simply normal humans breaking current safety rules.

12 Sporting Superhumans

Look at the situation one South African potential Olympic sprinter finds himself in. Oscar Pistorius competes using two carbon fibre lower legs made by the Icelandic bionic technology company Össur. Do these just return him to an exceptional level of sporting ability that he would have had, had he not been born deformed? Or are they actually enhancing him beyond what he would otherwise have had?

On 22 November 1986 Oscar Pistorius was born in Pretoria, South Africa. He was a healthy baby, but had a practical problem. Genetics had dealt him a slightly curious set of genes and one of the bones in each of his lower legs, the fibulae, was missing. When he was 11 months old, his doctors and parents decided that the best way forward was to amputate Oscar's legs halfway between his knees and ankles.

It was a traumatic decision, and the last thing that would have been on their minds at the time was the thought of seeing their son become a world-class athlete. But by 2007 he had acquired three world records in the double amputee class and two internationally recognised nicknames – 'the fastest man on no legs' and 'blade runner'. He competed against famous names such as six-time 100-metre world champion Marlon Shirley and the ever-competitive Brian Frasure.

Running on blades extends the capability of people with damaged limbs. So does a motorised car. The problem is deciding when one mechanical tool gives 'unfair' advantage.

By 2007 he had **acquired** three world records in the **double amputee** class

Shirley's and Frasure's Histories

Marlon Shirley was abandoned by his mother aged three, lost his left leg to a lawnmower at an orphanage aged five or so, got adopted at nine and has done well since. Brian Frasure was a college athlete with a taste for danger, and was hopping boxcars with his college friends when he missed a landing and got his leg trapped between two cars.

Oscar Pistorius won three gold medals at the 2006 IPC World Championships, in the 100, 200 and 400 metres. In the 200 and 400 metres, Oscar set new world records of 21.66 seconds over the 200 metres (achieved in the semi-final) and 49.42 seconds for the 400 metres.

There is no shortage of amputee competitors in sport today. The massive improvements in prosthesis technology in the last 15 years have enabled sprinters, long jumpers, high jumpers, golfers, skiers, snowboarders and even skateboarders to return to competition after losing body parts the rest of us consider essential. The feet made by College Park Industries, for example, and often fitted by Dale Perkins' prosthesis company Coyote Design are renowned for the careful design that makes them the most accurate replicas of fully functional fleshy human feet.

College Park's TruStep is worn by skateboarders like Garry Moore, champion surfers such as Sean Fitzgibbon, snowboarder Buddy Elias, motorcyclists like Chris Johnson (who was ranked third off-road trials rider in his state recently), and one tough martial arts fighter, Ronald Mann. Ron was initially turned down for a number of martial arts clubs because they were afraid his artificial leg could be used as a weapon. But he's since proved that his new leg doesn't make him any more dangerous than he already was. If you want to go fast, though, you need to talk to Össur.

Ron was initially **turned down** for a number of **martial arts** clubs because they were **afraid** his **artificial leg** could be used as a weapon

This Icelandic engineering company are not content to reproduce the behaviour of natural limbs. They are looking to go beyond. 'Life without limitations' is their mantra. Their racing feet, the unfortunately named Cheetahs, are made from layers of carbon fibre, like a racing car's body. And again unlike College Park's robust ankle joint with its carefully engineered degrees of motion and lifelike flexing, Össur just let one smooth, curved piece of carbon fibre do the job.

Cheetahs

Is the name a mistake, or a giveaway? If they enable the users to match the fastest land animal in the world, do they mean you are no longer performing as a human?

Or does using them break international rules and conventions, turning the user into a cheat?

The result is sheer speed and exhilaration for their users. And competitive success for those dedicated enough to do the training. Take British marathon runner Richard Whitehead for example. Born without legs below his knees, Richard started training for the New York marathon, his first ever, by running on his knees in his local gym back in Yorkshire.

'That's how I run when I don't have my prosthetic legs on. I then braved the roads, running mainly at night so as not to attract any attention! This was hard work, running in the dark. I found myself getting pains on top of other pains; some days I thought I was just running backwards. At that stage the most I could run for was about half an hour. But I needed to keep going, as November was getting closer and I hadn't run anywhere near five miles, so marathon distance seemed impossible.' He writes on his webpage on Össur's site.

Just 13 days before the marathon, still unable to do more than 10 miles, a friend's tip to explore Össur paid off, and Richard was fitted with a donated pair of Flex-Run feet, each one worth £2,000.

'My prosthetist advised me to take it easy and just wear these feet around the house for a couple of days. So I took to the road and ran for two hours. Oh, how that hurt. With two days to go I needed to push myself to breaking point, so with water and a banana I ran until I needed to get a lift home. I managed 18 miles, and I could not walk. I felt sick and distressed thinking I would struggle to make the journey, never mind the run.' Richard writes. On the day he finished in just 5 hours and 18 minutes and raised £8,000 for Macmillan cancer research.

Richard was entered with the disabled runners, but another Össur customer, American Mike Hicks, only ever competes against the able-bodied. Mike had been a very successful swimmer in his youth. But a car hit him while he was cycling one day and destroyed his left leg. Once over the initial healing he got back on his bike and took to competitive cycling. In 2006 he had three first places in his age group in the Mountain Bike Association of Arizona race series. He uses Modular III and Vari-Flex feet for riding and a selection of other Össur products for other activities. Not too controversial, great tales of perseverance against adversity.

Entering Competition

The problem is in deciding how much of these athletes' success is down to their power and personal dedication, and how much credit should be laid at the feet of the designers and engineers. Sorting it out is not easy, partly because everyone is unique and brings his or her own set of circumstances to the training circuit.

Perhaps losing a limb focuses the mind on a goal to win. Perhaps having to fight against prejudice and difficulty all your life makes you just that bit more determined to beat others. But perhaps the prostheses do as Össur claim and go beyond normal limits.

It certainly caused a stir when Oscar Pistorius' times began to out-compete other runners at the South African Olympic trial, and although he was allowed to run in two able-bodied events in the summer of 2007, the IAAF are still not sure whether to accept him as a competitor for the Olympics if he qualifies. They say that they want athletic competitions to be between human beings and not manufacturers; and also that unless you ban technical aids, what is there to stop someone showing up with rollerskates or a bicycle? The IAAF is saying that Oscar is welcome to compete in an appropriate event – the challenge for them is deciding which event is appropriate.

Össur engineer limbs and are keen to see their blades pounding the tracks in mainstream competitions such as the Olympics. They are also keen to show that their design leads to maximum performance. While flesh and blood limbs are multifunctional in that they are used for standing, walking, climbing, etc., Össur could potentially develop a limb for each sport. Such unifunctional limbs could have distinct advantage.

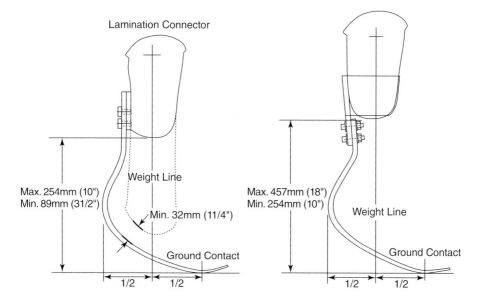

Out Classed

Like it or not, Oscar has found himself at the sharp end of a debate that has been running for years, and will run for years more. Who should compete against whom in sport? Should men and women have separate categories of competition, and different levels of prize money? Should unpaid amateurs have events from which paid professionals are barred? Should people who sprinkle steroids on their cereal each morning be allowed to play against others who just take sugar?

Should people who sprinkle steroids on their cereal each morning be allowed to play against others who just take sugar?

The route to the answer is via another question: why do sport in the first place? In athletics there are two possible answers. Do we watch the Olympics because we want to see someone win, or because we want to see records fall? Are we interested in seeing who can use their natural abilities to run fastest, or are we interested in a time-trial against the clock, an opportunity to see how fast we can make a human run?

It strikes me that we can have either, but probably not both at the same time. We have football clubs that divide their competitions into under-10s, under-12s and so on. We have men's and women's championships in tennis that run side-by-side in the same tournament. We have boxing competitions that regulate who meets who based on the competitors' weight. We have rowing regattas where there are numerous classes dependent on how geriatric you are. The list is endless.

In each case, the aim is to create fair competition, to generate a situation in which like is pitched with like, so that the winner is the person who does most with what they have been given. We are thrilled when our children do well in their age group, not because they have broken any world record, but because they have done well for who they are.

The Olympic games currently include 35 sports shared between summer and winter venues, but each of these is subdivided into many categories. Swimming is broken down into lengths of race and styles of stroke; skiing into the nature of the slope and whether or not there are obstacles to negotiate. All are divided into male and female competitions. In each case there are clear entry criteria and behaviour requirements for competitors. Behind this are desires for variety and fair play.

There are other occasions when individuals pit themselves against the clock, where they set out just to break a record, but even then the record is normally accompanied by a description: the fastest person to climb the Eiger unaided, or the first person to climb Everest without bottled oxygen.

Then again there are the records that are applauded because they are feats of technological achievement. The world land speed record doesn't record the power of human legs, but the combination of the inventive genius of a team both technically and in their ability to raise money and promote their project – oh, and a skilled driver with an unusual approach to risk who picks up the cup. Formula One motor racing recognises both technical teamwork and individual performance in that a prize is given each year to the winning driver, and another to the winning constructors, though again the playing field, or race track, is kept level by shed loads of rules accompanied by court-loads of lawyers.

The question facing Oscar, and any other sports man or woman who brings a novel piece of technology to the track, is whether he is now competing like with like. Is the fix wiping out previous disadvantages or creating a new set of advantages? In Oscar's case, the answer is not easy to find, because it is so difficult to discern what a like with like comparison should be. You could say that losing his lower legs and replacing them with lightweight alternatives will make him lighter, and therefore faster. The carbon fibre blades will also not be prone to lactic acid build-up and fatigue that plagues many an athlete. On the other hand his thigh muscles will have to do proportionately more work than other competitors'.

Many commentators have pointed to the length of his blades, saying that this gives him a longer than natural leg, enabling him to have a greater than normal stride. One consequence is that he has a slow start, but makes up for it once he is up to speed. The blades may make solid contact with the ground, but they have no capability of making the sorts of complex adjustments that a normal bodied person's ankle and foot can achieve. It means that keeping in lanes, particularly around a corner is a tricky feat. This lack of adjustment also means that he has greater difficulty in the wet, when the surface grip is reduced. This alone could cause problems for other competitors who don't want to be tripped or sliced by his blades if he leaves his lane.

Furthermore, muscles and tendons do a great job of absorbing the energy as a foot strikes the ground, and dissipating it. This saves joints from excessive damage and helps maintain balance. There is a springy nature in the muscles, tendons, cartilage and bone, and some of the energy stored can be fed back into the next step, but much is lost. One thing that is uncertain is quite how much of that energy is stored and returned in Oscar's limbs. The work has yet to be done. At first you might think that the manufacturers would be keen to stress the energy efficiency of their prosthesis – but right now they have gone to ground, and Oscar himself instead seems more anxious to talk about how much they waste.

A Run-in with the IAAF

In March 2007, as Oscar looked more competitive, the International Association of Athletics Federations (IAAF) amended its competition rules to ban the use of 'any technical device that incorporates springs, wheels or any other element that provides a user with an advantage over another athlete not using such a device'. It claimed that the amendment was not specifically aimed at Oscar, but they did start monitoring his track performances using high-definition cameras to determine whether he actually has an advantage.

The **amendment** was not specifically aimed at **Oscar**, but they did start monitoring his track **performances** using high-**definition** cameras

'With all due respect, we cannot accept something that provides advantages', said Elio Locatelli of Italy, the director of development for the IAAF, urging Oscar to concentrate on the Paralympics that will follow the Olympics in Beijing. 'It affects the purity of sport. Next will be another device where people can fly with something on their back.'

Oscar's coach, Ampie Louw, insists that the Cheetahs don't give Oscar an advantage. In fact he insists that he faces only disadvantages, as he has to work much harder to balance and corner and to make up for lacking the muscles of his lower legs. Louw added that Oscar's success is down to his general athletic ability.

What Happens When They Get Better?

Along with other commentators, regulators and observers of sport, the IAAF is faced with two issues when they watch Oscar run. First is the lack of knowledge about the human body and, as a consequence, an inability to come to firm conclusions about where to draw a line between a prosthetic replacement and an engineered enhancement. And secondly there is the realisation that technology is always coming up with new ideas, and will quite probably create devices that in any given niche can out-do the 'normal' human equivalent.

The argument against Oscar and his blades is one of a slippery slope – the issue of 'what's next?' The chances are that his current carbon fibre feet are just the first of what could soon become a long line of ever more capable prosthetic devices. According to Oscar his are basically a 14-year-old design. 'It's not like it is new technology, or unfair technology', he claims on a YouTube broadcast.

The idea is not lost on Össur. There is a second section to their website, labelled Bionic Technology, in which they talk about their future plans to seamlessly integrate their prosthetic technologies into 'an intimate extension of the human body' as Hugh Herr, the inventor of their Rheo Knee, informs us, with spooky intensity, at the close of the video introduction to the site. His vision is followed by a quote from George Bernard Shaw:

'Some men see things as they are and ask why, Others dream things that never were and ask why not.'

Here they promote not just the theoretical technologies they are now beginning to develop, such as osseointegration and neurosensing, but also products they are already testing on patients or selling, such as the Rheo Knee, the Proprio Foot and the Powerknee. These use electrical energy from lithium ion batteries, and clever electronic control systems to power artificial joints and make the action more natural, letting them respond to different terrain or inclines in the same way that undamaged human nerves and muscles can. This is technology that could easily lead to enhanced performance. A runner with legs based on this technology would be able to go on for as long as his batteries and moving parts lasted, but don't run too far from home – at the moment they only last 48 hours.

If we **ban** these **sorts of modifications**, **would** we also **ban someone who** has a steel pin inserted into her bone to aid strong **recovery** after a fracture?

Among ethicists, Oscar's success has spurred talk of 'transhumans' and 'cyborgs'. Some note that athletes already modify themselves in a number of ways, including baseball sluggers who undergo laser eye surgery to enhance their vision and pitchers who have elbows reconstructed using sturdier ligaments taken from elsewhere in the body. If we ban these sorts of modifications, would we also ban someone who has a steel pin inserted into her bone to aid strong recovery after a fracture?

On top of this, what happens when someone says, 'Why should I be limited to a stiff prosthesis when I am competing against someone who has a spring in their step when their natural feet hit the ground? Why shouldn't I incorporate springs?'

The effect of this would be dramatic, as anyone who has seen someone wearing spring-equipped stilts will know. After only a relatively small amount of training you can leap high in the air and run, at great speed, in straight lines. How would you feel if your basketball team found themselves pitted against people who could leap two or maybe three metres into the air before passing or catching the ball, and stride in two-metre paces down the court? The occasion would be a farce.

And if someone with missing feet or limbs can use springs, why can't a fully able-bodied person use them if his or her legs have stiffened up after injury or excessive training? If the answer is that you can only use the springy legs if you don't have legs of your own, then do we tempt athletes to enhance their chances of winning, to enhance their bodies, by amputating their limbs? The thought is bizarre, but it is a logical extension of taking a *laissez faire* approach to regulation in sport.

With minimal training, a basically fit person can leap way over any normal height – so long as he or she straps on springs.

REPRINTED WITH PERMISSION FROM SIMEON DIGNAM-CROTTY (WWW. RISERRAPTORS.COM)

Do we **tempt athletes** to enhance their chances of winning, to **enhance** their bodies, by amputating their limbs?

One possible answer is that you could set up rules that say you can compete as long as the spring coefficient of the limb is no greater than either the average of that of the current competitors, or perhaps the same as that of the currently most successful athlete or of the currently competing athlete with the greatest spring coefficient. It wouldn't be easy either deciding which option is fairest or taking the measurements.

But isn't that just the IAAF's point? They have tried to set limits to any bolt-on technology, but it is those very limits that people like Oscar and his helpers are complaining about. But all sports need rules and categories so that individuals and teams know what they are up against.

For those who would like to see the boundaries pushed, this would be disappointing. But there may be an opening. They look to a time when it will be at the Paralympics where sports records tumble, not the Olympics, and believe that when that happens, the focus of the world's attention will switch to this modified arena. It's a possibility, but I'm not convinced. No one denies that wheelchair marathon racers are powerful and dedicated when they complete the course in around one and a half hours, but that doesn't detract from the competitors who take half an hour longer and finish on their feet.

I've got nothing against having the world springy-stilt basketball championship, or holding rollerskating speed trials. For that matter I think it is a very good idea that people wear parachutes before trying to break world sky-diving records. But it seems a matter of common sense that sports work within categories, and at the moment I think that popular interest rests with the unmodified form. I believe that, at heart, we still hanker after the ancient Olympic idea of seeing the naked, or at least unaided, human strive and succeed. We still applaud those who stretch the unenhanced body.

Pure human competition means performing naked!

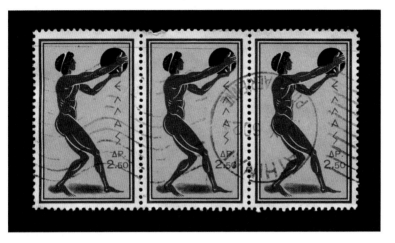

©ISTOCKPHOTO.COM/RISAMAY

We **still** hanker after the ancient Olympic idea of seeing the naked, or at least unaided, human **strive** and **succeed**

Military Mobility

How does Össur's technology fare away from sport? Jim Bonney is a Captain in the Royal Marine Commandos. His specialism is Landing Craft and recently he's been serving on HMS Ocean. Captain Bonney's right lower leg, ankle and foot were replaced with a prosthesis in December 2002.

In his journal, Jim Bonney tells how he fell around 1,000 feet onto a glacier while mountaineering in Alaska on a military climbing expedition. This broke both his ankles, and, while the left healed, the right did not, leaving him in increasingly unbearable pain during his first attempts at rehabilitation.

After meeting an ex-paratrooper with a similar injury who had chosen to have his leg amputated and was doing well with a prosthesis, Bonney decided to do the same. But it wasn't an easy choice and he tells with magnificent honesty what the consequences were for him, his wife and their young family. His self-esteem, career and income depended on being able to take his new leg to places no one had ever reached before on a prosthesis.

Bonney was determined to remain in active service, and has to pass the standard fitness tests. To complete the three-mile runs in 26 minutes in boots and trousers, he uses an Icex 100 socket, Iceross sleeve and pin, and a Ceterus Flex-Foot, all from Össur. He has a separate 'water leg' for his water-based duties and for kayaking for pleasure. He uses the Flex-Run (the same basic design as Oscar Pistorius' Cheetah) for running for pleasure and he's found that this is also ideal for two-man kayaking. It is narrow enough to get into the boat's cramped footwell and also great for the portages which have to be run while carrying the boat on competitive races such as the Devizes to Westminster Canoe race.

It is one up on changing your shoes to fit the task. He can change the engineering of his leg depending on the activity. It raises the possibility of building specialised prostheses that optimise efficiency in different activities with no need to put up with the sort of multi-use compromise found in natural limbs.

He can change the engineering of his leg depending on the activity

Bonney comments that falling off his mountain bike is less painful because the carbon fibre socket he wears takes the impact on that side. And he carries a spare leg and Ceterus foot just in case he has a major failure at a critical time. So these legs are not susceptible to some of the things that would injure a fleshy leg. And if you do fall and break it, you can just get the other one out, change it, which takes less than 90 seconds, and carry right on. That assumes, of course, that you have plenty of money and a rucksack of spares with you. Bonney wears out sockets – the carbon fibre parts that join his stump to the leg – at a rate of about one every six months. They cost £500 each. And after nine months the Ceterus foot is showing signs of terminal wear.

He carries a spare leg and Ceterus foot just in case

While the disposability of mechanical body parts has an instant allure as a means of coping with damage, it also points to the amazing qualities built in to our natural frame. Marines, for example, keep up a punishing schedule and they do manage to last more than six months before being retired for having terminally incapable joints. In our enthusiasm to be amazed at human engineering ingenuity, let's not forget to be doubly amazed at the built-in capabilities of the human body.

Although the legs and feet themselves seem to be well engineered, the join between the mechanical and the fleshy parts is a weak point. Legs come loose, sweat has to be dealt with chemically, and despite Captain Bonney's great discipline and care for his stump he's still suffered various infections.

Therapy or Enhancement?

Oscar Pistorius never ran on his own legs and so we can't make before and after comparisons between how great an athlete he would have been and the speeds he can achieve on his Cheetahs. We cannot tell whether they allow him to run faster than he otherwise would have done.

Captain Bonney, on the other hand, was already a Marine capable of passing the combat fitness test before he had his accident. And as he's only lost one leg, not both, there is more to compare. Although he is back on active service, and he did fully pass the fitness test, the Medical Board noted that Bonney was less able to carry heavy weights over long distances on difficult terrain because it put so much additional stress on the prosthesis. He would therefore be less effective than other colleagues on certain assignments. The Board consequently agreed to exempt him from duties such as arduous mountain treks in Afghanistan and assigned him to his favourite activity, small boat work.

Maybe he could get around this if someone designed a prosthesis that would be better suited for the heavyweight work. It would possibly have its own power to overcome the extra stress on his hip muscles, although that would then mean that the user would have to carry a powerpack or battery charger in his already bulging bag.

In Captain Bonney's case, so far, the leg is therapy. It has returned him to the level he was before his accident judging by his ability to pass the fitness test, and been robust enough to allow him to go on and fulfil some of the ambitions he had before the accident, such as competing in the canoe race. But it has weak points. Is it an enhancement?

Not yet, I think, but it has interesting potential.

V   Hype or Hope?

So what of my quest for evidence that human enhancement is here, needs to be taken seriously and should start to inform policy-making in the twenty-first century? It's been an interesting summer, and I've had the opportunity of meeting some notable characters. I've met some enthusiasts and people who believe that enhancement is a present reality and are keen to push enhancement forward as quickly as they can. But I remain a sceptic. I've encountered some strong and fascinating science and ground-breaking technology, but when it comes to human enhancement, a new super species, I am left believing there is more hype than hope – certainly at the moment and probably in the near future. If we are talking about enhancing human experience, creating enhanced lifestyles, then there is much more hope of a positive outcome.

In the Introduction of this book I started my journey by trying to establish a definition of enhancement. Naïvely I thought that this would give a solid foundation to my quest. I soon discovered that few people will restrict themselves to a single definition of enhancement, and when they do, it is often different from other people's definitions. That is frustrating because a central tenet of good communication is that words have meanings.

Few people will restrict themselves to a single definition of enhancement

To throw the net so wide as Todd Huffman did, and say that anything that makes life better than living in a technology-free cave is an enhancement, makes the term virtually meaningless. It becomes just an adjective to describe the human condition. We are a species that is marked by its persistent use of tools, and the desire to create societies that let groups of people collectively do more than the sum of the individuals. So yes, we are enhanced above a naked ape, but then that is nothing new, and needs no fresh description of differentiation from normal humanity.

To restrict it to saying that an enhanced human is one who has developed so far that he or she can no longer breed with the version of the species currently walking the planet moves it to such a high level of requirement that in most cases the answer is almost bound to be 'no'. The two main options for this level of enhancement are brain uploading and gene therapy. Both are far, far from being realised in any way that will lead to a super race. The transhumanist goal is therefore far from being achieved.

The in-between response is to see an enhanced human as someone who has adopted technology to the point that they have significantly greater abilities than any previous generation. In this regard I'd agree with Kevin Warwick. To call something an enhancement, the added feature should really add something new. It strikes me that most people who come to this area for the first time assume that by human enhancement we are talking of technologies that enable human experience to be heightened or intensified, extended or exaggerated. In addition they want to see the value, importance and attractiveness of people being increased.

I think it is still interesting to break the question into two categories. One looks at whether an individual is enhanced, the other at whether the technology creates an enhanced human community.

Evidence of Enhanced Individuals

The people whose lives have been most obviously enhanced by built-in technology were people like Fenella, the cochlear implant patient, and those who had received implanted brain electrodes. For them, novel applications of small electronic devices had been life-transforming – they could hear or be free from pain, depression or unwanted movements.

If we say that they are enhanced humans, we will need to go down a route of saying that everyone who receives a medical treatment is enhanced. After all, the implants haven't cured them, or given them abilities that are superior to those of 'normal' humans. The interventions have gone some way to correcting a loss or a deficit, and that is wonderful, but I'd be much more comfortable describing them as terrific therapies.

It is probably only a few years until someone builds a mobile phone into the cochlear implant, thereby making the person permanently on call. This would be getting closer to an enhanced individual. That is, it adds a new distinct built-in function, although I'd hate to be incapable of leaving the phone at home on a day off. Sticking electrodes in the brain also gives options for near-future enhancement. Turning on a 'high' for a party or flicking a switch to beat depression is new and could be a great tool and toy. It is not risk-free, as brain surgery should not be taken lightly, but society does have its risk takers, and the novelty is so great that I can see takers.

In terms of enhancement, it would introduce novel abilities, and quite possibly a novel level of existence. One issue that any person would need to be careful about would be ownership of the controller. You wouldn't want someone else dictating your mood at any given moment. That would soon turn it from enhancement to entrapment.

My biggest surprise came from visiting Peter Houghton. I'd assumed beforehand that he would see himself as an enhanced individual, but his insistence that he wasn't is a good illustration of the difference between therapy and enhancement. He is the first to say that he is a cyborg, a machine–human combination, but because the machine is not good enough to restore normal health much less improved performance, Peter refused to say that he was enhanced.

For him, just living longer is not necessarily enhancing. Nevertheless the assist is an amazing piece of engineering and one day a new generation of artificial heart will undoubtedly fill some of the gap. If it can pump at normal rates and automatically respond to the body's needs, plus have a manual override for the moments you would like to take direct control, then it would break through from therapy to enhancement – but that is a long way off.

Putting magnets in your fingers can seem a bit of a gimmick and I think Todd Huffman would be the first to admit that that is partly the case. It does, however, make the point that until you have got a new sensation, it is very difficult to know what it is going to be like to live with it. Todd's magnets and Kevin Warwick's electronic implants definitely give new ways of sensing the world, and both adventurers are convinced that the technology creates an enhancement. This is a particularly strong statement from Kevin as he reserves this term for something that is totally new and supra-human. Detecting a magnetic field by having your finger buzz as Todd does is intriguing, and until you have a new sensation you can never tell what it will do. I do think that it will be some time before there is a mass uptake of people who are willing to accept the risk of losing their finger for the fun of the new sensation, so don't believe that sensing magnet fields is going to be the turning point of human history.

Detecting a **magnetic field** by **having** your finger buzz is **intriguing**, and until you have a **new sensation** you can **never** tell what it will do

You could argue that people would be more likely to try a sonar-equipped helmet with stick-on buzzers, and simply put it on whenever they want to use it. This, however, would best be described as a tool. Warwick would argue that detaching it would also mean it was no longer you – and if so, you were no longer enhanced. I'd agree with that.

Looking at the technologies for enhancing life duration leaves me more sceptical. Aubrey's seven-point plan is interesting and the points that he wants to sort out are I believe all too real, but so too is his analysis that if you overcome six and leave a seventh unaddressed you will achieve little life extension. It would be like maintaining only six out of seven vital systems in a plane sending it back into the skies – the one unattended feature will still bring it down. It will be very hard to address all seven all the time in an individual, much less in all people.

In terms of life extension what we are seeing is more people reaching towards the top of the species maximum. That brings the average up, but I am not convinced that the upper limit is actually shifting. There have been a few long-lived people throughout history – what is happening is that more of us are getting towards that point. Similarly, any attempts at raising brain performance seem to hit an upper ceiling, with those who deliberately create memory strategies showing just how capable the trained brain is.

As for uploading, I think that this really takes the biscuit when it comes to hype. It can play a great role in science fiction, where the idea of a brain in a machine can let us explore questions about life, existence and what it is to be a sentient individual, but it is far removed from reality. The most enthusiastic people are the computer scientists like Anders Sandberg who really do believe that it is just around the corner. The least enthusiastic are the neurobiologists who see the complexity of the brain.

Where the computer scientists undoubtedly have more hope is when they set out to emulate a brain, or at least to emulate the thinking part; we can already see machines that can add fast, that can collect images and analyse them, looking for specific faces in the crowd in a way that no human could. We have financial computers that are so powerful they almost seem to have a life of their own. But they don't have a life of their own and would not make a good friend or soul-mate to relax with on holiday. Let's not get carried away by techno-hubris.

Deliberately forgetting may be more exciting. A technology that could selectively wipe out or at least dampen a person's response to trauma would enhance their life. The more complex ideas of wiping memory are more problematic because they would only work if you wiped everyone's brains at the same time, which is not likely to happen. In addition, enhancing memory beyond anything that humans are capable of is made awkward by the realisation that with training, the normal brain can do substantially more than most make it achieve. I'm also deeply sceptical that we will build a brain-databank interface in my lifetime.

When it comes to stretching our mind's ability to stay alert and focused on a task, we can certainly drop in some drugs and boost performance. One problem, however, is that any chemical you swallow, inhale or inject washes over the whole of your brain. The consequence is that it has a wide-ranging action on the nervous system, resulting in many unwanted symptoms. If you could deliver small packets to individual nerve terminals, things could be different.

People do things with **drugs** that **they** would never think of doing **without** them

On the other hand, even if the effects are a bit dirty and non-specific, people do things with drugs that they would never think of doing without them. They have ideas that no drug-free people have ever thought of before. They can stay awake for longer, concentrate harder and at times think more imaginatively – so isn't that enhanced? I remain uneasy. I'm not sure whether they are enhanced or screwed up.

I was excited by the way that you could take the tiny electrical signals that can be measured with a couple of electrodes stuck on a person's scalp, and detect brain waves. I'm amazed at the way that some people use it to drive computers, but I was disappointed to discover how far it had to go before it would do more than enhance the life of people who have severe movement disorders. It is very clever, and a superb tool for disabled living, but is not about to enhance humanity.

Finally, what of enhanced athletes? Drugs certainly do squeeze more power out of people's puff and place them at the limits of human capability. They can definitely enhance individual performances, but need using carefully if you plan to have a long life once the competition is over. Gene therapy is on the horizon and if it arrived would radically change the game, but right now there is nothing available for people in search of a boost – gene therapy is currently only an option for a few highly cosseted lab mice. All the same, WADA are probably right to make sure that they stay ahead of the game, monitor scientific progress and make sure they have thought out how to respond rapidly to developments.

So what of the enhanced individual? If your definition includes therapy, then yes there are plenty of examples, but we are not really discussing anything more than the technology and ethics of good medicine. If we are looking for something more exciting, more extending, then the immediate options are more limited.

Evidence of Enhanced Communities

We have a more interesting time when we ask whether applying technology to individuals ends up creating enhanced human communities. If we continue thinking about sport for a moment, there would only be any excitement in competition between enhanced people if all of the competitors were similarly enhanced. Without that you would have a situation akin to a heavyweight boxer taking on a fly-weight. Once you have matched the people enhancement for enhancement, the competitive advantage would be lost, and so would some of the appeal.

Some but not all. There is distinct excitement watching motorsports, sailing or aerial acrobatics, none of which would be possible without mechanical enhancement. They certainly draw in crowds who pay small fortunes for the privilege and then try to emulate their enhanced heroes as they drive home once the event has finished. The challenge will be in terms of defining appropriate categories for competition, but that is a legal more than a scientific or ethical issue.

Thinking again of Peter Houghton you could argue that although he personally didn't feel enhanced, the community has. By stopping him dying seven years ago it has helped to increase the average age expectancy of people in this country. Our collective life-expectancy has been enhanced.

But what of a whole community of people who had magnets in their fingers and could all sense magnetic fields? What would we create in our built environment to use that new sensation? Would we have labels in lifts that have magnetic sensitivity? And could you make sensitivity sufficiently discriminating to transfer that information? Would that be an enhanced human community? Would people without the implant start to miss out – start to become marginalised?

It is an important issue, but again is not a novel one. We are talking about a technologically advanced society. Now if you want to call that 'enhanced' then ok, but you need to be careful you don't imply it is a radical change from previous situations. Having a mobile phone has altered the way I live, and has led to a consequent loss of public phone boxes on streets and railway stations, so it has altered society. Would I say it had enhanced it? The answer to that question will be the same as the one for whether a society with magnetic labels is enhanced.

Having implanted identifying chips would become an issue if the next generation of passport included an implantable chip that linked you to your document. I wonder where you would put it? You would want to choose some part of the body that was common to everybody and easily accessed. The back of the hand would be good, but would discriminate against amputees. The same would be said for the top of an arm. Everyone has a chest, but not everyone want people reaching out at it with a chip reader. How about implanting it under the skin of the forehead with a little tattoo to show that you have been chipped?

A uniquely **identifying** code could **in one shot replace** all your credit and **debit** cards, your **club membership** and **loyalty** scheme cards, your library card and **meal** tokens

Just think of the advantage. A uniquely identifying code could in one shot replace all your credit and debit cards, your club membership and loyalty scheme cards, your library card and meal tokens. It could become your driving licence and tax badge. It could identify your remains if the majority of your body is destroyed in a violent assault or tragic accident. But it would only be of advantage if a large number of the population had it installed – maybe a forced decree that all babies had it inserted at the time of their first inoculation. It would bring with it the threat of enhanced identity theft – just think, if someone cloned your chip they would have stolen your wallet and filing cabinet where all personally identifying documents are stored. While I don't believe this would result in an enhanced individual, it would give rise to a new level of technologically advanced society and community. It is a community that would have a seriously powerful level of surveillance, and would create distress among civil liberty groups. I wonder, would the chip be a sign of enhancement or the mark of the Biblical beast?

In terms of public policy, the problem could become how we enable people who own different levels of technology to live side by side – the same issue as letting people without phones still get access to the dwindling number of pay phones. You would create a class or a clique of people with the enhancement. Only time would tell whether the benefit of having it would become so compelling that everyone else is forced down that line.

Would the chip be a sign of enhancement or the mark of the Biblical beast?

Public policy has an important issue with another therapy that sits on the border of enhancement, namely immunisation. A normal immune system responds powerfully to attacking bacteria or viruses that it has met before, but has a weaker response to new germs. Immunising someone lets their immune system size-up to a foe that it has never met. In some cases, once you have been given the jab your immune system is alerted for life.

Now, if you give one person the vaccine, you will achieve very little for the person or the community. There will be no reduction in the amount of circulating disease-giving organisms, and even the vaccinated person could receive a mild infection – if they happen not to have responded well to the immunisation, they could even have a serious bout of disease. Vaccination only has a major effect once 80 to 90% of the population has gone for it. That creates a community with so-called herd-immunity where the disease is most unwelcome, and soon heads off in retreat. Through acting on many individuals we have a modified, possibly even a resistance-enhanced community.

I could see a call from politicians to immunise our community from terrorism by injecting chips into people as an ultimate identity card. I'd be nervous about the campaign and, like many, would work hard to fight against any move towards compulsory chipping.

Pause for Policy

I love living in an advanced society with technology at hand, but I do so as a human, and I expect future generations to do so as well. I expect to be challenged by the arrival of new technologies, and at times people will be displaced from their employment by machines that do the job more efficiently. I am pleased to live in a country where we no longer send people underground to mine coal with picks and shovels, and am uneasy that other countries still exploit their labour forces. I expect to see a continuation of the trend in which big machines and information systems are driven by fewer people who individually have increasing power. I am not alone in pointing out how this will let the gulf between the 'haves' and 'have nots' widen as a consequence. But one thing that doesn't keep me awake at night is the fear of a superior species emerging from our inventions.

Is there any value in realising that a new superhuman race is a long way from occurring? Well I think so. Here is an extract from a paper written by Nick Bostrom and Anders Sandberg, the enhancement enthusiasts mentioned earlier in the book:

"Conventional" means of cognitive enhancement, such as education, mental techniques, neurological health, and external systems are largely accepted, while "unconventional" means – drugs, implants, direct brain-computer interfaces – tend to evoke moral and social concepts. However, the demarcation between these two categories is blurry. It might be the newness of the unconventional means, and the fact that they are currently still mostly experimental, which is responsible for their problematic status rather than any essential problem with the technologies themselves.

The paper makes interesting philosophical arguments, but it needs to be read with caution in terms of scientific reality. One of the reasons why people view these two categories differently is that education has a significant benefit, and brain implants don't work. Indeed Bostrom and Sandberg go on to point out that the best that technological methods can do in terms of cognitive enhancement is a 10% increase, while cognitive training can result in 1,000% gains.

Despite this we find them arguing in the end of the paper that we need to encourage open access to enhancements. 'The societal benefits of effective cognitive enhancement may even turn out to be so large and clear that it would be Pareto optimal to subsidize enhancement for the poor just as the state now subsidizes education', they conclude.

The desire to break away from human limitations seems to have clouded their vision. Despite being in possession of the best data showing that cognitive enhancement is very far from occurring, they are still calling for changes in policy to make it available to all.

Autonomy Hand in Hand with Regulation

You can see a similar argument in sports. In his 2004 book *Genetically Modified Athletes*, Andy Miah reviews the problems that sport faces in dealing with gene therapy. The book makes interesting reading, but before you throw away notions of ever seeing normal people run again, it is worth noting that only nine of its 178 pages of argument consider the state of the science. And those pages use some space pointing to current limitations. Despite that, Miah points to the convention of human rights that says people have a right to be treated equal irrespective of their genes, and says that sports will need to bow to the power that is driving change.

If we don't like the way that the rules of living are currently set out, fine, let's have a debate. If you want to alter the ethical values in society and sport, let's discuss it. But let's be careful to keep the discussions as real as possible.

I was surprised on this journey by one thing in particular. Much of the drive for enhancement comes from a school of thought that seeks to increase a person's ability to decide what want he or she to do with his or her life. The buzzword is autonomy. Anyone who attacks or even questions the sense of autonomy quickly finds himself or herself on the receiving end of ridicule at best. How then does this autonomy play out?

In sport we should apparently have the ability to choose what drugs, genes or physical enhancements we want – but it must be done in a frame of tight regulation. This regulation would then extend beyond those who sought enhancement to every member of the population so that the sports-police can set safety levels of features like hormones and ban anyone who steps over the lines. We can modify, but we will need to create new classes of athlete and make sure that the right people are in the right class. In drug taking, there is a call to liberalise, but this again needs to be kept under a watchful eye to make sure that the wrong people don't join in. An uploaded environment would be one where we would need incredibly tight management of cyber firewalls, and a powerful legal and practical framework that prevented hacking. There seems to be a contradiction here, as more rules lead to more regulation and hence less autonomy.

If deep brain stimulation ever got to the point of being fitted for fun, then we would need to ensure that no one could send out signals and take control of the population's mood or behaviour. Or we could decide that some people could use it for therapy, some for fun, and some instead of being locked away. The latter group would have their controller held by the some branch of the criminal justice system – a body that powerfully restricts autonomy. How about a chip that kept track of your every move? How much legislation do you think we would need before we decided how to regulate control of the information that system collected? It would certainly be a long way from a chip that gives freedom.

Now I've never been one to defend autonomy too passionately, because I believe that one person's actions almost invariably affect others. But those who do have a yearning to increase the power to take control over their lives should pause and see which way human enhancement technologies are likely to swing the balance between freedom and control.

Better Lives

We need to realise that there are plenty of opportunities for enabling people to live better lives. There are many ways to help individuals to reach up high and achieve at a level way above their current level of performance. There are increasingly large numbers of techniques and therapies that can transform the lives of people whose bodies are damaged. But few of these create a person who is enhanced beyond the maximum that has ever been seen in humans, and I would argue that none create a separate category of being that could be called an enhanced human.

Index